Simulation of the Morphogenesis of Open–porous Materials

Von der Fakultät Energie-, Verfahrens- und Biotechnik
der Universität Stuttgart zur Erlangung der Würde
eines Doktors der Ingenieurwissenschaften (Dr.-Ing.)
genehmigte Abhandlung

Vorgelegt von
Franz Keller
geboren in Immenstadt im Allgäu

Hauptberichter : Prof. Dr.-Ing. U. Nieken
Mitberichter   : Prof. Dr.-Ing. M. Piesche

Tag der mündlichen Prüfung:
21.11.2014

Institut für Chemische Verfahrenstechnik
der Universität Stuttgart
2014

D 93

Bibliographic information published by the Deutsche Nationalbibliothek

The Deutsche Nationalbibliothek lists this publication in the Deutsche
Nationalbibliografie; detailed bibliographic data are available
on the Internet at http://dnb.d-nb.de .

ISBN 978-3-8325-3962-7

Logos Verlag Berlin GmbH
Comeniushof, Gubener Str. 47,
10243 Berlin
Tel.: +49 (0)30 42 85 10 90
Fax: +49 (0)30 42 85 10 92
INTERNET: http://www.logos-verlag.de

Meinen Eltern

Vorwort

Die vorliegende Arbeit entstand im Rahmen meiner Tätigkeit als Wissenschaftlicher Mitarbeiter am Institut für Chemische Verfahrenstechnik (ICVT) der Universität Stuttgart. Mein herzlichster Dank gilt meinem Doktorvater Herrn Prof. Dr.-Ing. U. Nieken für seine wertvollen Anregungen, dem gewährten Freiheitraum und seinem allzeit entgegengebrachten Vertrauen.

Herrn Prof. Dr.-Ing. M. Piesche danke ich für sein Interesse an meiner Arbeit und für die freundliche Übernahme des Mitberichts.

Mein weiterer Dank gilt Herrn Prof. Dr.-Ing. G. Eigenberger für seine vielzähligen Ratschläge und seine Unterstützung bei der Verfassung der Forschungsanträge.

Allen Kollegen danke ich für die einmalige Arbeitsatmosphäre am ICVT. Durch das kameradschaftliche Verhältnis untereinander wird mir diese Zeit stets in bester Erinnerung bleiben. Besonders bedanke ich mich hierbei bei meinen SPH-Kollegen Winni und Manuel, den Kicker-Kollegen Stefan, Christian L., Christian S., Vanessa, Matthias und Carlos sowie bei Jens, Phillip, Karin und Katrin.

Der Deutschen Forschungsgemeinschaft danke ich für die finanzielle Förderung meiner Arbeit im Rahmen eines Normalverfahrens sowie durch den Sonderforschungsbereich 716.

Mein besonderer Dank gilt meinen Eltern, die mir durch ihre stete Unterstützung und ihr Vertrauen mein Studium und diese Arbeit erst ermöglicht haben. Von ganzem Herzen danke ich meiner Sarah für ihre Geduld und dass sie während dieser langen und arbeitsreichen Zeit immer an mich geglaubt und mir den Rücken frei gehalten hat.

Mannheim, Januar 2015   Franz Keller

Abstract

Open-porous functional materials play a crucial role in the field of chemical engineering, e.g. as catalysts, adsorbents and membranes. In many cases, the properties and performance characteristics of these materials are strongly determined and controlled by their open-porous morphology, which is specifically adjusted during the manufacturing step by directed manipulation or chemical modification. Currently, the development of production procedures for such open-porous materials relies almost completely on empirical correlations and experimental experience, while no established model based support of the development process is available.

In the present work, a contribution is made towards the development of a methodology for describing mesoscopic structure formation processes of open-porous materials. Based on the Smoothed Particle Hydrodynamics (SPH) framework, a meshless method has been developed, which is able to describe all of the relevant physical and chemical processes during the manufacturing step by solving the conservation equations of mass, momentum and energy on a quantitative basis. And since the governing equations are discretized using the particle based SPH formalism, the motion of the SPH particles is derived from the laws of continuum mechanics. Hence, no additional variables for parameterizing the inter-particle forces or stabilizing the numerical scheme have been introduced. Based on that and the Lagrangian nature of the SPH method, the developed approach is capable to accurately simulate the evolution of a heterogeneous multiphase body with evolving interfaces. In addition, large deformation as well as fragmentation of material with partly complex rheological behavior can be modeled. Moreover, the interaction of compressible and incompressible phases, the coalescence of voids as well as diffusion and chemical reaction in the deforming body can be described within the developed framework.

Furthermore, a strong focus has been laid on the accurate discretization of free surfaces and interfaces of the multiphase system within the developed framework. Therefore two corrected SPH schemes have been applied, which increase the accuracy of the SPH discretization at free boundaries and interfaces. Both approaches differ in the degree of correction. The moderately corrected approach, abbreviated by NSPH, is based on a renormalized kernel function and a simplified correction of the gradient operator, while the gradient operator is fully corrected by using a Taylor row expansion in the so called NCSPH approach. Additionally, a simplified hybrid approach for treating gaseous phases in the context of open-porous structure formation, consisting out of a particle based as well as grid-based part, has been developed in the present work.

The accuracy and numerical stability of all developed SPH approaches have verified at the hand of several test cases. While good agreement with the reference solutions has been achieved for the majority of the test cases for both approaches, the NCSPH method proved to be especially suited for the simulation of thin structures with predominating boundary effects. This is a very important feature of the NCSPH approach, making it especially suited for the simulation of open-porous structures. After validation of the developed approaches, a reaction-induced pore formation process by release of a blowing agent has been studied and the complete structure formation process by release of a blowing agent has been simulated using the NCSPH approach. Thereby, all characteristics of a pore forming process have been addressed in the simulation and the applicability of the developed SPH based methodology to describe the formation of open-porous materials on a quantitative level has been demonstrated.

Deutsche Zusammenfassung

Modellbildung und Simulation in der Verfahrenstechnik konzentrieren sich bisher haupt-sächlich auf die Beschreibung der makroskopischen Stoff- und Energieströme in ver-fahrenstechnischen Apparaten und Prozessen mittels Kontinuumsmodellen. Für die Ableitung der darin benötigten thermodynamischen Parameter und Transportkoeffizien-ten werden zunehmend auch molekulardynamische Simulationen eingesetzt. Dagegen ist das Zwischengebiet zwischen der molekularen Skala und der Makroskala einer detail-lierteren modellmäßigen Beschreibung und Simulation bisher kaum zugänglich, obwohl die Funktions- und Gebrauchseigenschaften vieler Materialien von ihrer mesoskopischen Struktur bestimmt werden. Diese Tatsache trifft besonders auf offenporige Funktionsma-terialien zu, die in vielen Bereichen der Technik eine wichtige Rolle spielen. So werden in der chemischen Prozess- und Verfahrenstechnik Katalysatoren und Adsorbentien benötigt, die eine wohl definierte Zugänglichkeit zu einer großen inneren Oberfläche mit spezifischen Oberflächeneigenschaften besitzen. Die Gebrauchseigenschaften dieser Materialien werden meist erst durch gezielte Beeinflussung oder chemische Modifikation der dabei gebildeten inneren und äußeren Oberflächen eingestellt.

Für die Fertigung von Funktionsmaterialien mit poröser Struktur existiert eine Viel-zahl von Herstellungsprozessen. Da entsprechende Modelle und Simulationsverfahren fehlen, basiert die technische Herstellung dieser Produkte bisher fast ausschließlich auf experimentellen Erfahrungswerten und empirischen Korrelationen. Dies führt u.a. bei der Maßstabsübertragung vom Labor- zum Produktionsprozess, bei der gezielten Veränderung der Morphologie oder der Änderung der Ausgangsmaterialien zu um-fangreichen zeit- und kostenintensiven Anpassungsexperimenten. Eine modellbasierte Unterstützung durch Vorhersage der zu erwartenden Materialstruktur und Morphologie in Abhängigkeit der Herstellungsbedingungen wäre daher sehr wünschenswert. Allerdings erweisen sich derartige Simulationen bisher als ausgesprochen schwierig, da die Morphologieausbildung durch Phasenumwandlungen, große Materialdeformation, hohe Dichteunterschiede zwischen Poren und Matrix, Fraktur von dünnen Stegen, Koaleszenz von Blasen sowie die Bildung interner und externer Grenzflächen in komplexen, sich stetig verändernden Geometrien charakterisiert ist. Daher wurde in den eigenen Vorarbeiten zunächst die Anwendbarkeit der gitterfreien Diskrete-Elemente-Methode (DEM) für die Simulation der Strukturbildung untersucht. Sie erlaubt eine qualitative Abbildung der bei der Strukturbildung auftretenden Phänomene. Allerdings müssen die erforderlichen Potenzialparameter zwischen wechselwirkenden Partikeln in der Regel empirisch durch Vergleich mit Referenzsimulationen angepasst werden. Daher wird in der vorliegenden Arbeit die Methode *Smoothed Particle Hydrodynamics* – SPH – aufgegriffen und deren Anwendbarkeit für die prädiktive Simulation der Morphologieausbildung von offenporigen Materialien untersucht. Im Vergleich zu DEM besitzt die SPH-Methode den großen Vorteil, dass die Bewegungsgleichungen der Partikel direkt aus den jeweiligen Kontinuumsmodellen abgeleitet werden können. Damit basiert die Simulation auf der etablierten Beschreibung aller relevanter physikalischer und chemischer Transport- und Umwandlungsprozesse, allerdings unter gitterfreien Bedingungen. Bevor die entwickelten Simulationsansätze genauer vorgestellt und deren Ergebnisse diskutiert werden, wird im folgenden Abschnitt das konkrete Anwendungsbeispiel kurz erläutert.

Wie bereits eingangs erwähnt, erfolgt die Herstellung offenporiger Funktionsmaterialien auf vielfältige Weise. In Rahmen dieser Arbeit beschränken wir uns auf die Erzeugung eines offenen Porensystems in einem viskoplastischen Substrat durch reaktionsinduzierte Freisetzung eines Treibmittels. In einem vorangegangenen Projekt wurde in Zusammenarbeit mit dem Institut für Kunststofftechnologie der Universität Stuttgart ein neuartiger Adsorbensmonolith für den Einsatz in der Druckwechseladsorption entwickelt. Der Rohling in Monolithform besteht dabei aus einem Polymer-Compound, in dem Zeolithpulver als das aktive Adsorbens und thermoplastisches Wachs als Gleit- und Treibmittel dispergiert ist. Dabei liegt das Wachs in der Polymermatrix in Form kleiner, mikrometergroßer Inseln vor. Zur Aktivierung muss der Rohling einem Porenbildungsprozess unterzogen werden, um die gewünschte offenporige Struktur mit Transportporen zu den Zeolithkristallen zu erhalten. Dazu reagiert eindiffundierender Sauerstoff mit den

Wachspartikeln zu einem Treibgas, das die Poren öffnet und eine offenporige Struktur hinterlässt. Durch gezielte Wahl der Betriebsbedingungen kann eine Zerstörung der makroskopischen Monolithgeometrie verhindert werden. Dieses Anwendungsbeispiel beinhaltet alle Charakteristika eines Porenbildungsprozesses, beginnend mit der Diffusion des Edukts Sauerstoff durch die heterogene Compound-Matrix und der lokalen Bildung des gasförmigen Treibmittels durch chemische Reaktion an den dispergierten Wachsinseln. Der zunehmende Gasdruck in den entstehenden Poren führt zu teilweise großen Materialdeformationen in dem viskoplastischen Substrat, die letztendlich in einer offenporigen Struktur münden.

Ziel dieser Arbeit ist, am Beispiel der Herstellung von offenporigen Funktionsmaterialien, eine Methodik für die prädiktive Simulation solcher komplexer Strukturbildungsprozesse zu entwickeln. Diese Methodik basiert auf Bulk-Materialparametern, Transportgleichungen für Energie, Masse und Impuls, sowie Geschwindigkeitsansätzen für chemische Reaktionen und soll in Zukunft zu einer effizienten Entwicklung maßgeschneiderter poröser Funktionsmaterialien beitragen und aufwendige Trial-and-Error Experimente unnötig machen. Um die notwendige Flexibilität bei der Behandlung von großen Materialdeformationen, dynamisch veränderlichen Grenzflächen und Materialrißen zu erreichen, basiert der gewählte Ansatz auf der gitterfreien Partikelmethode SPH. Hierbei liegt das Hauptaugenmerk wiederrum auf der Anwendung und Weiterentwicklung der SPH-Methode um eine quantitative Vorhersage der zu erwartenden offenporigen Materialstruktur unter den jeweiligen Herstellbedingungen zu ermöglichen. Neben der Implementierung der notwendigen physikalischen und chemischen Modelle, liegt ein Schwerpunkt auf der konsistenten Beschreibung und Diskretisierung von internen und externen Grenzflächen innerhalb des SPH-Konzepts.

Die Arbeit ist dabei wie folgt untergliedert:

- Kapitel 1 beschreibt die simulationstechnischen Schwierigkeiten bei der Beschreibung des Herstellungsprozesses von offenporigen Materialien. Weiter wird ein kurzer Überblick über technisch relevante Herstellprozesse von offenporigen Materialien gegeben und im Anschluss die Ausbildung des offenporigen Transportporensystems in dem ausgewählten Anwendungsbeispiel detailliert beschrieben. Im letzten Abschnitt des Kapitels wird der Stand der Technik bei der Simulation von derartigen Strukturbildungsvorgängen mit gitterbasierten Methoden sowie gitterfreien Partikelmethoden dargelegt. In diesem Rahmen wird zudem genauer auf die eigenen Vorarbeiten mit der Partikelmethode DEM[1] eingegangen und anhand dieser Ergebnisse die Wahl der gitterfreien SPH-Methode für die quantitative Simulation von Strukturbildungsvorgängen motiviert.

- Die Grundlagen der SPH-Methodik werden in Kapitel 2 ausführlich dargestellt, wobei der Schwerpunkt auf die für das Anwendungsgebiet wichtigen Aspekte des SPH-Verfahrens gelegt wird. Neben der Herleitung der unterschiedlichen SPH-Diskretisierungsschemata, wird weiter gezeigt, dass das allgemeine SPH-Verfahren an freien Oberflächen und Grenzflächen, wie sie bei der Ausbildung von offenporigen Materialien vorkommen, zu einer inkonsistenten numerischen Formulierung führt. Im Rahmen dieses Kapitels werden Maßnahmen zur Herstellung der Konsistenz an diesen freien Oberflachen diskutiert. Dies kann zum einen durch Rekonstruktion des fehlenden Teils der jeweiligen Interpolationsdomäne des betrachteten Randpartikels durch Spiegelung der Nachbarpartikel an der freien Oberfläche erfolgen. Zum anderen kann das klassische SPH-Verfahrens zu einem sogenannten *Corrected* SPH-Verfahren erweitert werden, wodurch durch die Renormalisierung der SPH-Kernfunktion und der Korrektor der SPH-Approximation basierend auf einer Taylorreihenentwicklung die Konsistenz des Verfahrens wiederhergestellt wird. Als Ergebnis bleibt festzuhalten, dass beide Verfahren prinzipiell geeignet sind, die Inkonsistenz an freien Oberflächen zu beheben, jedoch das erste Verfahren, der sogenannte Virtuelle Partikel Ansatz, besonders bei der Simulation von offenporigen Materialien mit einem naturgemäß hohen Anteil an freien Oberflächen zu einem drastisch erhöhten numerischen Aufwand führt. Weiter werden in diesem Abschnitt die beiden meist verwendeten Ansätze für die numerische Beschreibung von zeitabhängigen kompressiblen und inkompressiblem Strömungen mit Hilfe der SPH-Methode dargestellt. Die Beschreibung kompressibler bzw. schwach kompressibler Strömungen erfolgt zumeist mittels einer Prädiktor-Korrektor Zeitintegration, wobei der hydrostatische Druck zu den jeweiligen Zeitpunkten über eine Zustandsgleichung bestimmt wird. Diese Methode wird im folgenden *Weakly Compressible* SPH (WCSPH) genannt. Einem ähnlichen Prädiktor-Korrektor Schema folgt das sogenannte *Incompressible* SPH-Verfahren (ISPH), das auf einem

---

[1]Diese Vorarbeiten sind in Zusammenarbeit mit Blanka Ledvinkova und Juraj Kosek, Institute of Chemical Technology Prag, entstanden.

Projektionsverfahren basiert und die Inkompressibilität des Fluides durch das Lösen einer Poisson-Gleichung für den Druck und anschließender Korrektor der Partikelgeschwindigkeiten und -positionen gewährleistet wird.

- In Kapitel 3 wird das mathematische Modell sowie die entwickelten numerischen Lösungsansätze vorgestellt. Das mathematische Modell basiert auf den Erhaltungsgleichungen für Masse, Impuls und Energie, sowie den Komponentenmassenbilanzen. Daraus resultiert ein System gekoppelter, partieller Differentialgleichungen, das mit Hilfe der Konstitutivbeziehungen für die jeweiligen Materialien geschlossen wird. Für die Lösung dieses Gleichungssystems wird ein gitterfreier, auf der SPH-Methodik basierender Lösungsansatz entwickelt, welcher besonders für die Simulation der reaktionsinduzierten Strukturbildung von offenporigen Materialien durch Freisetzung eines Treibmittels geeignet ist. Dank der Lagrangeschen Natur der SPH-Methode und dem Design des Simulationsansatzes, ist die gegenwärtige Methode in der Lage den Reaktions-Diffusions-Prozess innerhalb des sich entwickelnden, heterogenen Materials mit komplexer Rheologie und großen Materialdeformationen, Fraktur von dünnen Stegen sowie Koaleszenz von Poren in quantiativer Weise zu beschreiben. Zudem ist durch die Kombination des WCSPH- und ISPH-Ansatzes die Simulation von Wechselwirkung zwischen kompressiblen und inkompressiblen Phasen möglich. Während die kinematische Grenzflächenbedingung zwischen diesen beiden Phasen aufgrund des Lagrangeschen Charakters der SPH-Methode intrinsisch erfüllt ist, muss die dynamische Grenzflächenbedingung in dem gegenwärtigen Ansatz über ein spezielles Verfahren erzwungen werden. Bei Wechselwirkungen über diese Phasengrenzen hinweg, interagieren beispielsweise Gasphasenpartikel, deren Druck über eine thermodynamische Zustandsgleichung bestimmt wird und somit einen absoluten Wert besitzt, und Flüssigphasenpartikel, deren Druck mit Hilfe der Projektionsmethode berechnet wird und somit nur einen relativen Druck darstellt.

Wie in Kapitel 2 dargestellt, führt die klassische SPH-Approximation in Bereichen in welcher die Interpolationsdomäne durch Ränder begrenzt wird, wie z.B. an freien Oberflächen und Grenzflächen, zu einem inkonsistenten numerischen Schema. Jedoch können gerade diese Bereiche bei die Simulation von offenporigen Materialien überwiegen. Aus diesem Grund werden in der vorliegenden Arbeit zwei unterschiedliche SPH-Diskretisierungsansätze verfolgt, die sich in der Art und dem Grad der Korrektur der SPH-Kernfunktionsglättung und deren Ableitungen unterscheiden. In dem ersten Ansatz, dem sogenannten *Normalized Corrected* SPH-Ansatz (NCSPH), wird die Konsistenz der Kernfunktionsglättung zumindest für konstante Funktionen durch die Einführung einer renormalisierten Kernfunktion wiederhergestellt. Zusätzlich wird die Ableitung der Kernfunktionsglättung korrigiert, womit der Gradient eines beliebigen linearen Vektorfelds exakt berechnet wird. Dank dieser Korrekturen kann die mathematische Randbedingung der Partiellen Differentialgleichung direkt auf das jeweilige Randpartikel aufgeprägt werden. Da diese Korrekturen jedoch den numerischen Aufwand des SPH-basierten Algorithmus erhöhen, wurde zusätzlich der teil-korrigierte *Normalized* SPH-Ansatz (NSPH) entwickelt um die Notwendigkeit der unterschiedlichen Korrekturen für den betrachteten Anwendungsfall zu untersuchen. Ähnlich wie in dem vorangegangen NCSPH-Ansatz wird auch im NSPH-Verfahren eine renormalisierte Kernfunktion bei der Kernfunktionsglättung verwendet. Jedoch wird deren Gradient nur in vereinfachter Weise korrigiert, wodurch der numerische Aufwand auf Kosten der Genauigkeit an freien Rändern reduziert wird. Neben der Rekonstruktion der Interpolationsdomäne an freien Oberflächen, wird die mathematische Randbedingung, besonders für die Druck Poisson-Gleichung des Projektionsverfahrens, innerhalb der NSPH-Methode durch einen vereinfachten Virtuellen Partikel Ansatz aufgeprägt. Beide Verfahren haben gemein, dass die Partikelinteraktionen direkt aus den Gesetzen der Kontinuumsmechanik abgeleitet werden und keine zusätzlichen, sozusagen künstlichen Kräfte zwischen den Partikeln verwendet werden müssen um die Stabilität des numerischen Schemas zu gewährleisten.

Weiter wird in Kapitel 3 ein vereinfachter Ansatz, die sogenannte hybride-SPH-Methode vorgestellt. Dieser wurde speziell für die Simulation der Herstellung von offenporigen Materialen durch Freisetzung eines Treibmittels entwickelt. Da die Deformation der Porenmatrix und somit die Ausbildung des porösen Systems hauptsächlich durch den statischen Druck in den Poren getrieben wird, wird die Kinematik der Gasphase in diesem Ansatz vernachlässigt. Somit wird nur die Porenmatrix mittels SPH Partikel diskretisiert und die Gasphase mit Hilfe eines unterlagerten festen Gitters aufgelöst, worüber sich der Verlauf des Drucks in den Poren berechnen lässt.

- Die Verifikation und Validierung der entwickelten SPH-Ansätze für die Simulation der Strukturbildung bei der Herstellung von offenporigen Materialien durch Freisetzen eines Treibmittels ist Inhalt des Kapitels 4. Dabei wird der Schwerpunkt auf Testfälle gelegt, welche zumindest einen bei der Strukturbildung ablaufenden Einzelprozess verifizieren. Kapitel 4 gliedert sich dabei in drei Abschnitte. Im ersten Abschnitt wird der grundlegende, auf einer Projektionsmethode basierende SPH-Algorithmus verifiziert. Ein besonderes Augenmerk wird dabei auf die Rekonstruktion der Interpolationsdomäne an freien Oberflächen gelegt. Dabei hat sich gezeigt, dass der moderat korrigierte NSPH-Ansatz im Vergleich zu dem ISPH-Ansatz nach Stand der Technik im Allgemeinen genauere Ergebnisse liefert. Zudem wird die Stabilität des numerischen Verfahrens erhöht. Eine weitere Verbesserung der Genauigkeit sowie der Stabilität wird durch den NCSPH Algorithmus erreicht. Durch die Renormalisierung der Kernfunktion sowie der Korrektor der Ableitung der Kernfunktionsglättung können auch dünne Strukturen mit großen Randeinflüssen in stabiler und genauer Weise simuliert werden. Dies ist besonders für den konkreten Anwendungsfall, der Ausbildung einer offenporigen Struktur, von größter Wichtigkeit. Gerade für diese Art von Strukturen hat sich der NSPH- sowie der konventionelle ISPH-Ansatz nach Stand der Technik als ungeeignet erwiesen.

Im zweiten Abschnitt des Kapitels 4 wird die Eignung der entwickelten Simulationsansätze für die Beschreibung von komplexem Materialverhalten anhand von einfachen Testfällen untersucht. Für die Simulation des Materialverhaltens elastischer Festkörper erwiesen sich beide Ansätze als nahezu gleich geeignet. Dies gilt jedoch nur unter der Einschränkung, dass im Verlaufe der Simulation der elastische Festkörper nur kleinen Materialdeformation unterworfen ist. Da dies in dem betrachteten Anwendungsfall zutrifft, kann auf weitere Maßnahmen zur Erhöhung der Stabilität des numerischen Schemas bei großen Deformationen verzichtet werden. Ebenso sind beide Ansätze für die akkurate Beschreibung von viskosem Materialverhalten geeignet. Hierbei liefert der vollständig korrigierte

NCSPH-Ansatz leicht bessere Ergebnisse. Dies liegt in der Behandlung der Haft-bedingung des viskosen Mediums an festen Wänden begründet. Während die festen Wände innerhalb des NSPH-Ansatzes ähnlich wie an freien Oberflächen rekonstruiert und diesen virtuellen Partikeln Geschwindigkeitswerte mittels Punkt-spiegelung zugewiesen werden müssen, kann die Haftrandbedingung innerhalb des NCSPH-Ansatzes direkt auf die Randpartikel aufgeprägt werden. Diese ein-fachere Behandlung der Haftrandbedingung führt auch bei der Simulation der Strömung eines viskoplastischen Fluides zu leicht besseren Ergebnissen. Der Vor-teil der NCSPH-Methode bei der Behandlung der Haftreibungsbedingung kommt bei der Beschreibung von viskoelastichem Materialien dagegen voll zum tragen. Für die Beschreibung der Strömung von vollelastischen Fluiden nach dem Upper-Convected-Maxwell Modell ist der NSPH-Ansatz aus diesem Grund nicht geeignet. Für diesen Fall ist der NCSPH-Ansatz die Methode der Wahl.

Die Verifizierung des entwickelten SPH-Simulationsansatzes für kompressible–inkompressible Mehrphasenströmungen ist Bestandteil des letzten Abschnitts. Zunächst wird die Eignung des rein partikelbasierten SPH-Ansatzes für die Be-schreibung von Systemen mit sehr hohen Dichtedifferenzen zwischen den Phasen anhand eines statischen Testfalls untersucht und mit der analytischen Lösung verifiziert. Weiter wird der Aufstieg einer Gasblase in einer ruhenden, Newtonschen Flüssigkeit mit hohen Dichte- und Viskositätsdifferenzen zwischen den Phasen simuliert und der Verlauf des Blasenaufstiegs mit Ergebnissen eines gitterbasierten Verfahrens aus der Literatur verglichen. In beiden Fällen ergibt sich eine sehr gute Übereinstimmung mit der Referenzlösung. In einem weiteren Testfall, der Expansion einer kompressiblen Gasblase, beschrieben nach dem idealen Gasge-setz, in einer inkompressiblen Newtonschen Flüssigkeit, wird der entwickelte rein partikelbasierte Kopplungsansatz zwischen kompressibler Gasphase und inkom-pressibler Flüssigphase verifiziert. Gute Übereinstimmung wurde dabei mit dem NSPH- sowie NCSPH-Ansatz erzielt, wohingegen sich die klassische ISPH-Methode als nicht geeignet erwiesen hat. Zudem wurde der vereinfachte hybride-SPH-Ansatz mit Hilfe dieses Testfalls verifiziert. Unabhängig von der gewählten Diskretisie-rungsart wurde eine zufriedenstellende Übereinstimmung mit der analytischen Lösung gefunden.

- Die Simulation der Strukturbildung bei der Herstellung von offenporigen Materialien durch Freisetzung eines Treibmittels durch chemische Reaktion beinhaltet Kapitel 5. Zunächst wird die Simulation der Bildung und Öffnung einer Einzelpore mit Hilfe des rein partikelbasieren NCSPH-Ansatzes gezeigt. Aufgrund der durchweg sehr guten Verifizierungs- und Validierungsergebnisse des NCSPH-Ansatzes, wurden diese Simulationsergebnisse als Referenz für die weiteren Simulationsansätze gewählt. Dabei hat sich herausgestellt, dass der NSPH-Ansatz nicht für die Simulation der Strukturbildung von offenporigen Materialien geeignet ist, da die Strukturentwicklung bei beiden Methoden drastisch unterschiedlich verläuft. Wie in Kapitel 5 ausführend dargelegt, wird die deutliche Abweichung zwischen den beiden erhaltenen Strukturen auf die Inkonsistenz der NSPH-Diskretisierung des Gradientenoperators zurückgeführt. Die Simulation der Entstehung der Einzelpore mit dem entwickelten hybriden–NCSPH-Ansatz wird im Anschluß gezeigt. Dabei ist bis zur endgültigen Öffnung der Pore eine sehr gute Übereinstimmung zwischen dem rein partikelbasierten und dem hybriden Ansatz zu erkennen. Ein Abweichung in der Materialstruktur tritt gegen Ende der Simulation mit dem Beginn der Fraktur der Porenmatrix auf und wurde großteils der unterschiedlichen Behandlung der Gasphase zugeschrieben. Basierend auf dieser Analyse werden zukünftige Verbesserungspotentiale aufgezeigt. Darüber hinaus wird der gesamte Prozess der Ausbildung eines offenporigen Transportporensystems mit Hilfe des NCSPH-Verfahrens simuliert. Im Rahmen dieser Simulation werden alle in Kapitel 1 beschriebenen Merkmale des betrachteten Porenbildungsprozesses abgebildet. Anhand einer Parameterstudie kann die Auswirkung unterschiedlicher Polymer-, Zeolith- und Wachseigenschaften auf die endgültige Struktur aufgezeigt werden. Die Diskussion des Einfluss der Wachsphasenverteilung im Compound auf die finale Struktur schließt das Kapitel ab. Somit steht eine verifizierte Simulationsmethodik für die prädiktive Simulation der Strukturbildung während der Herstellung von offenporigen Materialien zur Verfügung.

- Kapitel 6 beginnt mit einer Zusammenfassung der vorliegenden Arbeit. Im Anschluss werden die aus Sicht des Autors notwendigen nächsten Schritte skizziert. Dabei ist vor allem die Erweiterung des entwickelten Simulationscodes auf große Teilchenzahlen und dreidimensionale Strukturen von herausragender Bedeutung.

Das Kapitel endet mit einem Ausblick auf die weiteren Anwendungspotentiale der entwickelten partikelbasierten Sᴘʜ-Methode.

Contents

CONTENTS

List of Figures

List of Tables

Glossary

MW_k molar weight of component k

N number of (neighboring) particles

n particle number density

p pressure

R universal gas constant

r_h reaction rate of reaction h

r_{cut} kernel cutoff radius

s specific entropy

T temperature

u specific internal energy

V particle volume

w weight function or kernel

Roman Symbols

A	matrix
b	right hand side
f	body force vector per unit mass
g	gravity vector
j_k	diffusive mass flux of component k
v	velocity vector
x	position vector
$\dot{q}$	specific heat flux
$\hat{n}$	normal unit vector
$\hat{t}, \hat{s}$	tangential unit vector
υ	specific volume
$\widehat{w}$	renormalized kernel
$\widehat{w}$	renormalized weight function
b_k^M	mobility parameter of component k
C_1, C_2, T_r	WLF constants
c_v	specific heat at constant volume
E	Young's modulus
H	height of domain
h	smoothing length
L	Lagrangian operator length of domain
m	particle mass

Greek Symbols

α	thermal diffusivity
β	retardation ratio
β^S	surface coefficient
β_T	isothermal compressibility
σ	total stress tensor
τ	deviatoric stress tensor
χ	color function
δ	Dirac delta function
$\dot{\epsilon}$	strain rate tensor
$\dot{\gamma}$	local magnitude shear rate
η	dynamic viscosity
Γ	denominator within NSPH
κ	local curvature
λ	thermal conductivity, 1. Lamé constant
λ_1	time constant of relaxation
λ_2	time constant of retardation

μ	shear modulus,
	2. Lamé constant
ν	Poisson ratio
ν_{kh}	stoichiometric coefficient
ω	angular velocity
ω_k	mass fraction of component k
$\overline{\lambda}$	1. Lamé constant in 2D
σ	surface tension coefficient
ϱ	mass density
ϵ	strain tensor
Ω	rotation tensor
τ_0	yield stress

Superscripts

α, β, γ	tensor index
$*$	intermediate time

Subscripts

0	initial state
i, j, k	particle index

Other Symbols

∇	derivative
$\widetilde{\nabla}$	corrected derivative
$:$	double-dot scalar product
$\otimes$	dyadic product

Acronyms

CSF	Continuum surface force
CSPM	Corrected Smoothed Particle Method
DEM	Discrete Element Method
DPD	Dissipative Particle Dynamics
DSPH	Discontinuous Smoothed Particle Hydrodynamics
FDM	Finite Difference Method
FEM	Finite Element Method
ISPH	Incompressible Smoothed Particle Hydrodynamics
LBM	Lattice-Boltzmann Method
NCSPH	Normalized Corrected Smoothed Particle Hydrodynamics
NSPH	Normalized Smoothed Particle Hydrodynamics
SEM	Scanning electron microscope
SPH	Smoothed Particle Hydrodynamics
VOF	Volume Of Fluid
WCSPH	Weakly Compressible Smoothed Particle Hydrodynamics
XSPH	Extended Smoothed Particle Hydrodynamics

1

Introduction

Open-porous functional materials play a crucial role in the field of chemical engineering. Typical examples are catalytic and adsorptive processes, where materials with a large specific internal surface with congruent accessibility are needed [1, 2]. For other functional materials, the open-porous structure is inevitable for the selective separation of fluids by steric effects or surface forces [3]. Presently, the development of manufacturing procedures for such open-porous materials relies almost completely on experimental experience and empirical correlations. Therewith, a model based support of the development process is highly desired. However, with the morphogenesis of open-porous materials being characterized by large material deformations, complex geometries, as well as fragmentation of material and coalescence of voids, the structure formation process itself is extremely challenging to describe nummericaly. And up to now, neither a methodology nor a validated numerical solver is available in the literature. In the present thesis, a contribution is made towards the development of such a method. In order to tackle the numerical difficulties, the particle based simulation technique Smoothed Particle Hydrodynamics (SPH) has been taken as a starting point. Due to its meshless character, no computational grid is needed to discretize the domain and the structure evolution is described by the motion of a set of particles. With it, the transport equations for energy, mass and momentum are solved together with rate expressions for the chemical reaction in a Lagrangian frame. In the present work, the method has been developed further and specialized for the simulation of the reaction-induced morphogenesis of open-porous materials.

1

This chapter gives a short overview over some technically relevant manufacturing processes of open-porous materials. Furthermore, the regarded application example is presented in detail, with a strong focus on the challenges arising during the modeling and simulation of the latter. Subsequently, possibly suitable simulation techniques for the desired application are discussed and an overview of the state-of-the-art for material structure formation is given in section 1.3.

1.1 Motivation

Simulations of macroscopic spatial distributions of concentration, velocity, temperature and pressure fields in a specified macroscopic environment are well established in the field of chemical engineering. On a molecular level, simulations are increasingly used to determine thermodynamic properties as well as transport parameters. In contrast, the so called mesoscopic scale between the molecular and macroscopic level (in the range of several nanometer to several microns) is insufficiently developed from a modeling and simulation methodology perspective. However, the internal structure formation process during the manufacturing of open-porous solids evolves on the mesoscopic length scale. And the properties of open-porous functional materials are specifically adjusted by directed manipulation or chemical modification of the evolving internal and external interfaces. Examples thereof are hollow fiber membranes, which are used as artificial kidneys in the dialysis process.

Since appropriate models and simulation approaches are missing, the technical production of these materials relies almost completely on experimental experience and empirical correlations [3]. Consequently, the scale up from laboratory to the production process, the directed modification of the morphology or the change of raw materials leads to further time-consuming and cost-intensive follow-up experiments. Hence, a model based support of the design process by prediction of the resulting material structure would be extremely helpful. So far, this kind of simulations prove to be notably difficult, since the morphogenesis is characterized by phase changes, large material deformations, high density differences between pores and pore matrix, fracture of thin ligaments, coalescence of voids, as well as continuously evolving internal and external interfaces in complex geometries.

Thus, for the predictive simulation of the morphogenesis of open-porous materials, a detailed and dynamic simulation of all relevant physical and chemical processes is inevitable. According to the author's knowledge, no established and validated general methodology for such structure formation process has been developed so far. Therewith, the presented work aims at the development of such a methodology to enable the simulation of complex structure formation processes for functional materials. Thereby, the developed methodology will contribute to an efficient and rational design of tailor-made open-porous functional materials and prevent elaborate trial-and-error experimentation. As a long-term goal, the work at hand will help to fulfill the vision of a simulation based integrated process and material development.

In the following section, a short overview over selected technically relevant manufacturing processes of open-porous materials is given to illustrate the general importance and applicability of the desired methodology. Subsequently, the regarded application example is presented in detail, with a strong focus on the challenges arising during the modeling and simulation of the latter.

1.2 Application examples

Numerous procedures for the production of porous materials have been developed. Examples thereof are the formation of open-porous materials by phase inversion. A particular illustrative example of the formation of an asymmetric precipitation membrane is shown in figure 1.1. Beginning on the left, water is added as precipitating agent on top of a polymer solution. With increasing time, the polymer solution segregates and a solid open-porous membrane is formed by precipitation as it is shown from the left to the right picture in figure 1.1 [3].

Figure 1.1: Photographs of the formation of an asymmetric precipitation membrane at different time instants. From left to right: Start of the developing interface after addition of the precipitating agent (water) on top of the polymer solution; porous structure after 12 s; after 24 s and final open-porous membrane after 5 min. Taken from [3].

Another example of a production process for porous solid particles is the spray drying or spray polymerization technique [4–7]. In this case, a liquid or slurry is dispersed by a spray nozzle or atomizer and dried rapidly in a co- or counter-current hot gas stream. In case of the spray polymerization process, the drying is supplemented by polymerization reactions. In this way, various open-porous particle morphologies can be obtained depending on the process conditions as shown in figure 1.2, where selected particles morphologies manufactured by reactive spray polymerization are shown [8]. Besides material composition, strict requirements are posed towards the particle morphology to ensure the processability of these solid particles. Thus, the particle porosity, texture and surface roughness needs to be adjusted in the process.

The common ground of the presented examples lies in the fact, that the material properties and performance characteristics are strongly determined and controlled by the open-porous material morphology. In addition, the structure formation proceeds on a mesoscopic scale during the manufacturing process. For this reason, the specific and systematic manipulation as well as chemical modification of the developing internal and

4

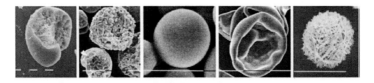

Figure 1.2: Scanning electron microscope pictures of the various particle morphologies observed by spray polymerization under different process conditions. Taken from [8].

external interfaces is highly desirable from a chemical engineering point of view. Another example falling into these categories, and where the focus is laid in the context of this thesis, are structure formation processes by formation and release of a blowing agent inside of a viscoplastic substrate. This will serve as an example for a pore formation process, which is characterized by large deformations, rupture of material, coalescence of voids and formation of new surfaces. However, the developed simulation approach is formulated on a general basis and not limited to the following application case.

Model system: reaction induced formation of an open-porous material

In cooperation with the Institute for Polymer Technology of the University of Stuttgart a novel adsorbent monolith with good adsorptive properties and minimal pressure drop has been developed for the application in rapid pressure swing adsorption (RPSA) processes in previous works [9, 10]. In contrast to the conventional approach, in which the monolithic structure is fabricated with an organic plasticizing aid and an inorganic binder [11], the backbone of the new monolith material consists of a polyamide polymer matrix. The polyamide backbones, which makes up to $20 - 35\,\mathrm{wt.\%}$, is highly filled with zeolite powder $(55 - 70\,\mathrm{wt.\%})$. Moreover, a thermoplastic wax is dispersed as internal lubricant and blowing agent in the compound $(5 - 20\,\mathrm{wt.\%})$ [9, 12]. The monolithic adsorbent structure, as it is shown in figure 1.3, is manufactured in a two-stage process. In the first step, the mixture is extruded in a twin screw extruder in a monolithic form.

As the zeolite crystals are fully encapsulated in the polymer, the material is not suitable for the RPSA process, where fast adsorption kinetics and rapid gas transport to the zeolite crystals are essential. Therefore, the monolith has to be processed further to generate an open-porous structure in a second step. After extensive experimental trials,

Figure 1.3: Typical adsorbent-polymer compound monolith geometry. Taken from [9].

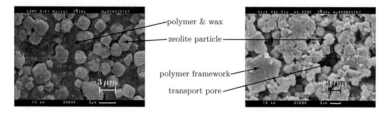

Figure 1.4: SEM images of an adsorbent-polymer compound monolith before (left picture) and after (right picture) thermal treatment. Taken from [10].

pore formation could be achieved by thermal treatment of the material in an oven under a reduced oxygen atmosphere. In figure 1.4, scanning electron microscope (SEM) pictures of the compound material structure prior to the pore forming step (left) and afterwards (right) are shown. On both pictures, the zeolite particles, which are in the range of some microns, can be easily identified. The thermoplastic wax is not mixed on a molecular level, but distributed in the compound in form of several micrometer large unconnected islands [13]. After removal of the wax by the thermal treatment at approximately $200\,^{\circ}C$, the picture changes drastically. The wax with a low molecular weight decomposes to form a blowing agent, which opens the pores and leaves an open-porous structure behind and the polyamide framework remains to ensure sufficient stability. The resulting macro pore sizes in the polymer compound are in the range of $0.8\,\mu m$ to $5\,\mu m$ [10]. Experiments on a differential scanning calorimetry apparatus showed, that the pore formation is based on a slow oxidation of the wax by the ambient oxygen permeating through the compound matrix. To prevent the polymer matrix from melting, thermal degradation or even burning, the temperature, residence time and oxygen partial pressure have to be adjusted carefully [9]. Furthermore, the amount of wax is a key parameter. Obviously, a too low concentration in the compound leads to a system with distributed closed pores. On the contrary, a too high amount of wax results in a initially percolating system. Therewith, the wax phase is not acting as a blowing agent, but is just burnt without widening pores. The optimal amount of wax for fast kinetics is found to be approximately 5 wt.% with a zelolite loading of 75 wt.% [9, 14].

The presented process has all characteristics for a pore forming process. A blowing agent is formed at distributed locations in the bulk material by diffusion of oxygen through the polymer matrix and the evolving pore system. The blowing agent causes large material deformation and generates a very open-porous structure. In the following section, an overview of the suited simulation techniques is given and the state-of-the-art in modeling open-porous structure formation processes is reviewed.

1.3 State-of-the-art in modeling open-porous structure formation

1.3.1 Comparison of numerical methods

As stated in the previous section, the formation of an open-porous structure by release of a blowing agent inside a viscoplastic fluid is investigated in this manuscript. The process leading to the deformation of the initial substrate is mainly characterized by evolving free surfaces and by rupture or coalescence of material elements. For the selection of an appropriate method for this type of problem, the different classes of mathematic methods are discussed shortly and specific advantages and disadvantages thereof are reviewed. In the author's opinion, commonly used mesh-based methods are not optimal for modeling the complex interplay of the above mentioned processes. This is due to the fact, that the arising large deformations are difficult to simulate even with advanced Finite Element Methods (FEM), like Arbitrary Lagrangian Eulerian FEM or Extended FEM [15–18]. Secondly, evolving free surfaces, edge fusion of voids and especially fracture and break-up are also tough to simulate by using sophisticated surface

7

reconstructions methods like the *Level Set* and *Volume Of Fluid* (VOF) method, which are needed for mesh-based Eulerian methods. A detailed discussion of the advantages, disadvantages and constraints of mesh-based methods and the surface capturing or tracking methods can be found in Appendix A.

By using particle methods, mesh constraints are eliminated, enabling the simulation of very large deformations. Furthermore, the need for tracking surfaces is eliminated. The advantages of particle methods in general, as well as the mathematical background of the most promising particle methods are summarized in the following section. A more detailed discussion of advantages and disadvantages of the particle methods *Dissipative Particle Dynamics* (DPD) and *Lattice Boltzmann Method* (LBM) for the application at hand is given in appendix B.

1.3.2 Mesh-free methods

Mesh-free methods are characterized by a set of particles, which carry all the information about the dynamic and thermodynamic properties of the material, such as position, velocity, stress, mass and density to name a few. These particles move according to external forces as well as forces induced by the whole interacting system. And by making the mesh constraints obsolete, very large deformations of a continuum can be handled easily, since the connection pattern among the particles is generated as part of the computation and changes over time. In addition, the temporal evolution of particle connections allows the handling of damage and fracture of materials in a straightforward way. This feature is essential for modeling the evolution of open-porous materials. Free surfaces can be handled easily and naturally without special emphasis on surface tracking as well. The particles move to new locations and the shape of the free surface is defined by the position of these boundary particles. Explicit surface tracking, as required for grid based methods, is obsolete. While surface tracking methods like VOF are well established for free surfaces, the problem becomes rather challenging if rupture of material or edge fusion of cavities takes place [19]. As open-pore formation is characterized by these processes, mesh-free methods are more suitable in this case. By basing the equations of motion on a Lagrangian formulation, accurate tracking of complex topological surface changes can be easily achieved. Moreover, additional and complex physical interactions, like non-Newtonian rheology can be included in a consistent way [20]. A further advantage of Lagrangian method is the fact, that each particle has its specific identity and intrinsic variables. These internal variables stay with the particles during computation, allowing one to include additional features like chemical reactions depending on concentrations in a simple way [21].

Although there are advantages, there are also general problems related to particle methods. The most important one is the treatment of fixed boundaries, like confining walls, which does not come naturally. Care has to be taken to prevent particles from penetrating the walls by applying special techniques, like reflecting the velocities at solid boundaries in order to model no-slip boundary conditions [22–24]. Also the modeling of in- and outflows into or from confined regions is challenging [25]. Despite the fact, that particle methods are gridless, there is still a form of meshing required to ensure that the spatial extent of the body is represented by the particles correctly. Hence, the particles must be initially distributed in a certain configuration. As stated by Swegle, this process is considered less difficult due to fewer constraints on the particle connection pattern in comparison to mesh generation in grid based methods [26].

While the properties presented above can be attributed to particle methods in general, particle based methods can be further classified into deterministic and probabilistic methods. Representatives of the latter are molecular dynamics [27], direct simulation Monte Carlo [27, 28], as well as Lattice Gas Automata and its derivatives [29]. These methods represent macroscopic properties as statistical behavior of microscopic particles. However, due to the very small length and time scales covered by these methods, they are computationally expensive, even for modern supercomputers, and can therefore not be applied directly to engineering problems at the mesoscopic scale. These methods are more suitable for understanding the fundamental atomistic interactions that underlie macroscopic materials, than for modeling of mesoscopic structure formation dynamics.

Grid-based methods for structure formation processes

The formation of closed cell foams has been investigated by several scientific groups with different numerical methods. Coupez and coworker simulated the expansion of a closed cell foam structure via unstructured finite elements using their commercial simulation code REM3D[1]. In all their works, the germ formation and nucleation is

---

[1]http://www.transvalor.com/rem3d_gb.php, date:10.10.2012

neglected and a previously defined number of germs is purport in the viscous substrate. In the Eulerian reference frame, the bubble growth has to be captured via interface capturing methods. Coupez *et al.* is using the interface capturing methods VOF [30] as well as the Level Set method [31, 32]. A similar approach has been chosen by Thuerey for the simulation of the expansion of a closed cell foam using the Lattice-Boltzmann method [33]. When using the LBM method, an explicit interface capturing method has to be used also [34].

As stated earlier, structure formation processes are characterized by large material deformations. If using Lagrangian gird-based methods, these large deformations can lead to strong mesh distortions, which can result in deterioration of accuracy or even stop the computation. As stated in the work of Li and Liu, these problems are in parts alleviated by using more advanced FEM approaches as stated previously. However, even with these improvements, large material deformations can still lead to errors, not only via remeshing of the computational domain, but also via the interpolation of the state variables on the new grid [35, 36].

Particle methods for structure formation processes

In contrast, the advantage of meshfree particle simulation methods lies in the absence of mesh constraints. Therewith, the particles can move independently in space and interfaces are developed quasi automatically via particle accretion or secession. While interface tracking or capturing methods are needed in multiphase simulations for grid-based methods, the additional computational expense is omitted for particle methods, since the respective phases are described by the respective particle types. This is not only an advantage for the above presented foaming processes, but in general for very heterogeneous materials. A further advantage of particle methods in comparison to mesh-based approaches lies in the relative simple treatment of material fracture.

A widely used approach for modeling macroscopic material behavior using particle methods is the so called Discrete Element Method (DEM) initially developed by Cundall and Stack [37]. So far, DEM is mostly used for modeling the dynamics of granular systems. Nonetheless, Grof and Kosek used the DEM methodology with enhanced material models to simulate the growth and structure formation of porous polyolefin particles in a catalytic gas phase polymerization [38, 39]. In [38] Grof *et al.* develop an approach for modeling reaction and transport processes in dynamically evolving multiphase materials, with which he is capable of describing the formation of cavities and fracture in the catalyst support and polymer phase. In the developed DEM approach, the monomer and polymer phase as well as the catalyst support and the generated pores are discretized by particles of different type. Chemical reaction, transport phenomena and mechanical deformation are computed on the interconnected particle network. An important point is the description of material deformation via inter-particle forces. Elastic deformations obeying Hooks law are reproduced by spring elements, whose forces are proportional to the inter-particle distance, while viscous material behavior is modeled via dashpot elements. Complex material behavior is approximated by combination of the latter elements, e.g. to the well-known Maxwell-model [40]. The solid polymer particles are growing at the cost of the surrounding monomer particles. Therewith, the inter-particle forces are altered, which leads to stress and tension between the particles, ultimately resulting in particle motion. With the presented method, Grof *et al.* is able to simulate the formation of macro cavities and hollow particles as well as catalyst fragmentation [41]. For the first time, the pore forming process in the polyolefin gas phase polymerization could be described on a qualitative basis. However, with the presented approach it is not possible to use the established continuum model parameters for the description of the rheological material behavior. In fact, the DEM spring-dashpot elements have to be parameterized more or less empirically. The parameterization can be done by comparison of DEM simulation results with the results of continuum models under well defined conditions, e.g. like the Couette flow test case. Therein, the decisive disadvantage of the DEM approach is seen for the predictive modeling of material structure formation, since available continuum models cannot be used directly and the validity range of the new particle interaction models remains unclear.

Modeling of the pore forming process using DEM

In a first approach, the discrete element method developed by Grof and Kosek and presented in section 1.3.2 was adapted for the modeling of the morphogenesis of open-porous materials by release of a blowing agent as shown in section 1.2. A detailed description of the joint work with Juraj Kosek and Blanka Ledvinkova is published in [42] and will be summarized in the following.

11

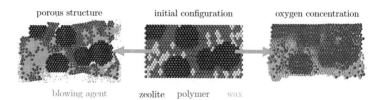

Figure 1.5: Left: Formation of a porous structure using the DEM approach; Center: Discretization of the adsorbent monolith using DEM particles; Right: Diffusion of educt oxygen in the pore system and matrix [42].

The computational domain, an excerpt of the polymer-zeolite compound, is shown in the center of figure 1.5. Zeolite particles are depicted by black particles, while the polymer backbone is represented by red particles. The wax phase is sketched by yellow particles and the resulting gaseous decomposition products are illustrated by green particles. In dependence of the respective particle interactions, the force interaction models are chosen. For interactions between two zeolite particles or two gaseous particles the purely elastic spring model is used. Polymer – polymer as well as wax – wax interactions are characterized by the viscoelastic Kelvin-Voigt material model [40]. As aforesaid, the development of the porous system is caused by the decomposition of the wax phase, the resulting pressure increase in the pore and the plastic material deformation, which ultimately results in local fracture of the polymer matrix. Thereby, the ongoing decomposition reaction is dependent on the local oxygen concentration. The decomposition process itself is modeled by shrinking wax particles. Simultaneously, gaseous particles are inserted at the surface of the wax particles, which grow over time under the constraint of mass conservation. In figure 1.5 the resulting porous structure at a certain instant in time is shown on the left. On the right side of figure 1.5, the corresponding distribution of the oxygen partial pressure is shown.

To conclude, the formation of an open-porous structure can be reproduced in a physically meaningfully manner using the presented DEM approach. However, several weaknesses of DEM for the present application have been identified. As reported in detail in [42], the repulsive and attractive parameters of the DEM material models, as well as the maximum strain between two interacting particles have proved to be determining for the resulting morphology. But these parameters cannot be derived from the macroscopic continuum mechanics parameters and need to be parameterized by simple reference simulations. Therewith, the validity of the DEM particle interaction models for non-parameterized applications is unknown and the validity of the DEM results in general is questioned. These results have been the main reason for searching for a methodology capable of solving the continuum equations on a quantitative level. Therefore, the particle interaction models have to be derived from the laws of continuum mechanics. In the following chapter, the foundations of the method Smoothed Particle Hydrodynamics (SPH) are presented. Based on the SPH formalism, the equations of motions of each particle are derived from the laws of continuum mechanics. This is considered as one of the most important steps towards the predictive simulation of material structure formation processes. Though beforehand, a SPH-based simulation of a structure formation process published in the literature, which exhibits some resemblance with the application example at hand, is sketched in the following section.

Modeling of the pyrolysis of uranium oxide using SPH

Before going into the details of the SPH formalism, one of the rare examples in the literature for the simulation of a structure formation process using the SPH method is sketched. A model for describing the pyrolysis during the fabrication of uranium-ceramic nuclear fuel has been developed and published by Wang and coworkers [43]. As initial condition, uranium oxide U_3O_8 particles are dispersed in a silicon carbide SiC matrix as shown in figure 1.6 on the left. Subsequently, these U_3O_8 particles react with SiC in a consecutive reaction first to UO_2 and afterwards to UC. During the conversion, the particles shrink and a gap between the SiC matrix and the UC particles is formed as sketched on the right of figure 1.6. On first sight, the published structure formation process has some resemblance with the one investigated in this work. However, the SiC matrix particles are fixed in space and hence, no deformation of the matrix is considered. Furthermore, a momentum balance is not solved in the presented work and the gap between the SiC matrix and UC particles is formed due to the shrinkage of the filler particles and a not otherwise specified pressure force only. In addition, it is stated by the authors, that the evolving gap is usually very small in the regarded case with a thickness of less than 25 % of the smoothing length, the characteristic length scale of the SPH interpolation. Assuming a typical smoothing length of 1.5 times the particle size, the gap is half as thick as the surounding particles. Nevertheless, the SPH approach

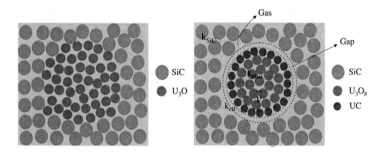

Figure 1.6: Models of the particle shrinkage and movement for the pyrolysis during the fabrication of uranium-ceramic nuclear fuel. On the left, the initial configuration of the SiC matrix and the filler particles is shown, while the gap formation and shrinkage is sketched on the right. Taken from [43].

proved to be easily applicable for the descripition of the strongly heterogenous material. In the following chapter, the fundamental principles of the SPH approach are presented.

2

The Smoothed Particle
Hydrodynamics Method

Due to its Lagrangian nature, the Smoothed Particle Hydrodynamics (SPH) concept seems to be a promising method for the simulation of structure formation process. The ongoing continuous evolution of internal and external interfaces, as well as fragmentation or fracture of material is generally not handled easily by standard grid-based methods. In addition, the equations of motion of the particles are derived from the laws of continuum mechanics using the Smoothed Particle Hydrodynamics formalism. Based on that, the thermodynamic properties of the material can be computed along the particle trajectory. Therewith, the conservation equations can be solved on a quantitative basis, enabling the predictive simulation of structure formation processes. In the present chapter, the basics of the SPH concept are presented. A special focus is laid on the construction of a consistent numerical scheme to enable an accurate treatment of interfaces and free surfaces within the SPH framework.

2.1 Basic Smoothed Particle formulations

Similar to the method of lines discretization in grid-based methods, a system of coupled, partial differential equations of continuous field quantities has to be transformed to a system of coupled, ordinary differential equations, whose values are evaluated at discrete spatial positions. In a first step, the position dependent functions are convoluted with a kernel function. Second, the convolution integral is approximated by numerical quadrature, using the particle positions as evaluation points. In the following section, these two aspects, which give rise to the name *Smoothed Particle Hydrodynamics*, namely, the smoothing of discrete function values by convolution with a kernel function and the discretization of the convolution integral using a particle approximation, will be presented. The information in this section is meant to provide an overview of the concepts of Smoothed Particle Hydrodynamics important for the application to structure formation processes and by no means a complete discussion of the topic. Several text books [35, 44–48] and review articles [20, 36, 49–52] are available which provide a more widespread and complete description of the theoretical aspects of SPH, its further developments as well as meshfree methods in general.

2.1.1   Kernel approximation and kernel function

The foundations of SPH are derived from the context of integral interpolant theory [53, 54]. Any linear interpolation operator can be written as integral representation of a continuous field function $f(x)$, yielding the following identity

$$f(x) = \int_{\Omega} f(x')\delta(x - x')d\,x', \tag{2.1}$$

with x being the position vector, $\delta(x - x')$ the Dirac delta function centered at position x and $f(x)$ the interpolated function defined on the entire domain Ω. In order to establish a discrete numerical scheme, the Dirac delta function $\delta(x - x')$ is replaced by the kernel or smoothing function $w(x - x', h)$. The scaling variable h possesses the unit length and is called smoothing length. The smoothing length h defines the extent of the kernel and accordingly the domain of its influence. Hence, the integral estimate $\langle f(x) \rangle$ is obtained as follows

$$\langle f(x) \rangle = \int_{\Omega} f(x')w(x - x', h)d\,x'. \tag{2.2}$$

The kernel estimate or kernel interpolation can be seen as a smoothing process, with function $f(x)$ being averaged over the entire domain, weighted by the kernel function $w(x - x', h)$. As stated above, equation (2.2) is an approximation of $f(x)$ as long as the kernel function $w(x - x', h)$ is not equal to the Dirac delta. Consequently, the kernel function has to obey several conditions [55, 56]. To ensure consistency of the interpolation in the continuum limit, the kernel function has to tend towards the delta function with the smoothing length h going to zero

$$\lim_{h \to 0} w(\boldsymbol{x} - \boldsymbol{x}', h) = \delta(\boldsymbol{x} - \boldsymbol{x}'). \tag{2.3}$$

To assure an accurate interpolation, the kernel needs to satisfy the normalization condition

$$\int_\Omega w(\boldsymbol{x} - \boldsymbol{x}', h) d\,\boldsymbol{x}' = 1. \tag{2.4}$$

In general, a spherically symmetric and even kernel function is chosen, which is only dependent on the distance of the interpolation points. With the definitions

$$\boldsymbol{r} = \boldsymbol{x}' - \boldsymbol{x} \qquad \text{and} \qquad r = |\boldsymbol{r}| \tag{2.5}$$

the following statements are valid:

$$w(\boldsymbol{x} - \boldsymbol{x}', h) = w(\boldsymbol{x}' - \boldsymbol{x}, h) = w(\boldsymbol{r}, h) = w(r, h). \tag{2.6}$$

In addition, the kernel function should have a compact support, resulting in

$$w(\boldsymbol{x} - \boldsymbol{x}', h) = 0 \quad \text{for} \quad |\boldsymbol{x} - \boldsymbol{x}'| > \alpha\,h, \tag{2.7}$$

where h is denoted as smoothing length of the kernel function and α is a constant related to that chosen kernel as it will be shown later in section 2.1.4. By using kernel functions with a compact support, the integration of the complete domain Ω is reduced to an integration over the localized support domain as shown in figure 2.1.

The accuracy of the kernel approximation shown in equation (2.2) has been investigated extensively in the literature [49, 56–61] and is found to be second order in space. By making use of the fact, that $w(\boldsymbol{x} - \boldsymbol{x}', h)$ is a strongly peaked function to preserve the local character of the averaging, $f(\boldsymbol{x}')$ can be expanded in a Taylor series approximation around $\boldsymbol{x}$. Since $w(\boldsymbol{x} - \boldsymbol{x}', h)$ is chosen to be an even function, the term of order $O(h)$ in the smoothing length vanishes, yielding the following expansion as exemplary derived by Laguna [62]:

$$\langle f(\boldsymbol{x}) \rangle = f(\boldsymbol{x}) + \frac{h^2}{2} \nabla^2 f \int_\Omega w(\boldsymbol{x} - \boldsymbol{x}', h)|\boldsymbol{x}' - \boldsymbol{x}|^2 d\,\boldsymbol{x}' + O(h^4). \tag{2.8}$$

17

With the second term been unequal to zero, the error estimate for $\langle f \rangle$ results in

$$\langle f(r) \rangle = f(r) + O(h^2), \qquad (2.9)$$

showing the second order accuracy of the kernel estimate $\langle f \rangle$ with respect to the smoothing length h. According to Speith, the kernel approximation shown in equation (2.2) is therewith exact for constant and linear functions [61]. However, these considerations are only true for full support of the interpolation domain and a detailed consideration of the kernel approximation for open-porous materials is given in section 2.3.

2.1.2 Discretization of the kernel approximation

Figure 2.1: Neighboring particles contributing to the kernel interpolation with compact support and cutoff radius r_{cut}.

In the second step of the SPH discretization procedure, the integral estimate of equation (2.2) has to be evaluated. This has to be done numerically, since function f is only known at discrete spatial points. To apply the interpolation to an arbitrary material, the material has to be divided into a set of particles with mass $\rho(\boldsymbol{x}')\,d\,\boldsymbol{x}'$. These particles are used as nodes for the numerical quadrature of the kernel interpolation. Since the particles, respectively integration nodes, are randomly distributed in space, the quadrature approach is called Monte-Carlo integration and will be sketched in the following [49, 61].

In order to obtain a discrete representation of the integral, an ensemble of N particles is randomly distributed inside the domain Ω as shown in figure 2.1. With known function values $f(\boldsymbol{x}_j)$ at the particle position $\boldsymbol{x}_j$ the particle number density n (particles per volume with the unit $\frac{1}{m^3}$ in the three dimensional case) is introduced for the derivation as

$$n(\boldsymbol{x}) = \sum_{j}^{N} \delta(\boldsymbol{x} - \boldsymbol{x}_j). \qquad (2.10)$$

According to Laguna, the consistency condition of the kernel function (2.3) can be expressed as [62]

$$\lim_{h \to 0} \langle n(\boldsymbol{x}') \rangle = \lim_{h \to 0} \sum_{j}^{N} w(\boldsymbol{x}' - \boldsymbol{x}_j, h) = n(\boldsymbol{x}'), \qquad (2.11)$$

resulting in the limit of $n(\boldsymbol{x}')/\langle n(\boldsymbol{x}') \rangle = 1 + O(h^2)$. Multiplying the kernel approximation in equation (2.2) with $n(\boldsymbol{x}')/\langle n(\boldsymbol{x}') \rangle$, yields the numerical integration of the kernel interpolation

$$\langle \langle f(\boldsymbol{x}) \rangle \rangle = \sum_{j}^{N} \frac{f(\boldsymbol{x}_j)}{\langle n(\boldsymbol{x}_j) \rangle} w(\boldsymbol{x} - \boldsymbol{x}_j, h) \qquad (2.12)$$

with the particle number density being the inverse of the particle volume $d\,\boldsymbol{x}' = 1/n(\boldsymbol{x}')$. With the introduction of the particle mass

$$m_j = \frac{\rho_j}{\langle n_j \rangle} \qquad (2.13)$$

one yields the commonly known SPH equation for an arbitrary function f at particle position $\boldsymbol{x}_i$

$$\langle \langle f(\boldsymbol{x}_i) \rangle \rangle = \sum_{j}^{N} \frac{m_j}{\rho_j} f(\boldsymbol{x}_j) w(r_{ij}, h), \qquad (2.14)$$

with r_{ij} being the distance between particle i and j. To point out the difference between the discretized SPH function value $\langle \langle f \rangle \rangle$ with the integral estimate of equation (2.2), the double brackets $\langle \langle \rangle \rangle$ have been used. With the presented formalism, which is called particle approximation, the integration of the kernel approximation has been replaced by the summation over discrete function values (including its own function value), weighted by the smoothing function. Therewith, particle interactions and attributes are calculated in the absence of a mesh topology, which can be seen as the key advantage of meshless methods in comparison to grid-based approaches, since the interpolation nodes can be randomly distributed in space and change their positions over time freely. In the successive part of this work, the abbreviation $f_j = f(\boldsymbol{x}_j)$ for a function value of the particle at position $\boldsymbol{x}_j$ is used. So, the latter equation is simplified to

$$\langle \langle f_i \rangle \rangle = \sum_{j}^{N} \frac{m_j}{\rho_j} f_j w(r_{ij}, h). \qquad (2.15)$$

Regrading the derived equation, it is obvious that the SPH particles are not mass points in the sense of statistical mechanics, but have a finite extension and can be seen as discrete parts of the body on the one side and as interpolation nodes on the other side.

In section 2.1.1, the approximation error of the kernel interpolation is estimated using a Taylor row expansion. In contrast, no (closed-form) expression for the discretization error of the numerical quadrature can be derived, since the error is dependent the on particle distribution [62]. The statistical error due to the Monte-Carlo integration is estimated to $\propto 1/\sqrt{N_p}$ under the presumption of N_p randomly distributed nodes [49, 61]. However, rather than a random particle distribution, some kind of regular structure is expected due to mutual particle interactions following the conservation laws as shown by Monaghan [49, 63]. In this case of moderate disorder, Niedereiter estimated the error as being proportional to $\propto N_p^{-1} \log N_p^{d-1}$, with d being the number of dimensions [64]. A slightly lower average error has been estimated by Wozniakowski [65]. Nevertheless, the discretization error can be reduced by increasing the number of particles, as shown in the works of Monaghan [56] and Martin *et al.* [66]. By choosing appropriate particle numbers N_p and smoothing length h, the discretization error can be effectively reduced. In addition, Laguna estimated the error numerically, by checking the normalization condition of the kernel function (equation (2.4)) in its discretized form

$$\sum_{j}^{N} \frac{m_j}{\rho_j} w(r_{ij}, h) = 1 \qquad (2.16)$$

for all particles. For bulk particles, i.e. particles with a complete particle neighborhood, the deviation from unity is attributed to an irregular particle distribution. This has been verified by Colagrossi, who reconfirmed, that the integral interpolation converges with second order for bulk particles distributed in a regular way. Contrariwise, the interpolation does not converge for an irregular particle distribution [67]. This observation has be refined by Quinlan *et al.* who demonstrated, that second order convergence is retained for irregular distributed particles, whose spatial perturbations from the regular and equispaced configuration are small with respect to the smoothing length h [68].

To conclude, the error in SPH simulations originates from two sources, as already stressed using the $\langle\langle \rangle\rangle$ notation. First, an error is introduced by kernel interpolation as shown in section 2.1.1. The second error source lies in the numerical quadrature of the convolution integral shown in equation (2.2), which is mostly dependent on the particle distribution.

2.1.3 Spatial derivatives of the kernel interpolation

Based on the formalism presented in the sections 2.1.1 and 2.1.2 above, the spatial derivative of a distributed function $\nabla f(\boldsymbol{x})$ can also be derived within the SPH formalism in two steps [61]. At first, the kernel approximation of the derivative $\langle \nabla f \rangle$ is obtained as defined in equation (2.2) by folding the derivative of the function ∇f with the kernel function $w(\boldsymbol{x} - \boldsymbol{x}', h)$. This procedure results in

$$\langle \nabla \cdot f(\boldsymbol{x}) \rangle = \int_{\Omega} w(\boldsymbol{x} - \boldsymbol{x}', h)\, \nabla' \cdot f(\boldsymbol{x}')\, d\,\boldsymbol{x}', \qquad (2.17)$$

where ∇' denotes the directional derivative along the direction of $\boldsymbol{x}'$. By applying integration by parts

$$w(\boldsymbol{x} - \boldsymbol{x}', h)\nabla' \cdot f(\boldsymbol{x}') = \nabla' \cdot \left[f(\boldsymbol{x}')\, w(\boldsymbol{x} - \boldsymbol{x}', h) \right] - f(\boldsymbol{x}') \cdot \nabla' w(\boldsymbol{x} - \boldsymbol{x}', h) \qquad (2.18)$$

and substituting the integrand in equation (2.17), the spatial derivative can be transfered to the kernel function, following in

$$\langle \nabla \cdot f(\boldsymbol{x}) \rangle = \int_{\Omega} \left\{ \nabla' \cdot \left[f(\boldsymbol{x}') w(\boldsymbol{x} - \boldsymbol{x}', h) \right] - f(\boldsymbol{x}') \cdot \nabla' w(\boldsymbol{x} - \boldsymbol{x}', h) \right\} d\,\boldsymbol{x}'. \qquad (2.19)$$

The first term in the volume integral on the right side of equation (2.19) can be replaced by an integral over the surface S using Gauss theorem. This results in

$$\begin{aligned} \langle \nabla \cdot f(\boldsymbol{x}) \rangle = \int_{S} f(\boldsymbol{x}')\, w(\boldsymbol{x} - \boldsymbol{x}', h) \cdot \hat{n}\, d\,\boldsymbol{S}' \\ - \int_{\Omega} f(\boldsymbol{x}') \cdot \nabla' w(\boldsymbol{x} - \boldsymbol{x}', h)\, d\,\boldsymbol{x}', \end{aligned} \qquad (2.20)$$

with $\hat{n}$ being the outwards-pointing unit normal vector. If a kernel function with compact support is used, the value of the kernel function $w(\boldsymbol{x} - \boldsymbol{x}', h)$ is zero at the outer boundaries of the integration domain as shown in figure 2.2. Therefore, the surface integral in equation (2.20) is zero for interior particles, respectively particles with full particle neighborhood. Though, in regions where the support domain is truncated by the computational domain, e.g. on free surfaces, that condition is not intrinsically satisfied. However, for free surface flows for example, the surface integral needs to be zero at the free surface, if the common stress-free boundary condition is applied. In that case, no special care has to be taken. Hence, if field variables are not zero on the surface, modifications have to be made to account for residual boundary effects. Since

21

boundaries play an important role in the desired applications, a deeper discussion will be presented later in section 2.3. For interior particles, the integral estimate of the first derivative results in

$$\langle \nabla \cdot f(\boldsymbol{x}) \rangle = -\int_\Omega f(\boldsymbol{x}') \cdot \nabla' w(\boldsymbol{x} - \boldsymbol{x}', h) d\boldsymbol{x}'. \tag{2.21}$$

With the presented formalism, the construction of a differential operator of a function can be achieved by (analytical) differentiation of the given kernel function, instead of the derivation of the function itself. Due to the anti-symmetry of the directional derivative of the kernel function, exemplary shown in figure 2.3a, equation (2.21) can also be expressed as

$$\langle \nabla \cdot f(\boldsymbol{x}) \rangle = \int_\Omega f(\boldsymbol{x}') \cdot \nabla w(\boldsymbol{x} - \boldsymbol{x}', h) d\boldsymbol{x}'. \tag{2.22}$$

The particle approximation of the latter equation, is obtained by solving the convolution integral using numerical quadrature with the particle being the interpolation nodes in the same way as presented in the section 2.1.2. Therewith, the SPH discretized first order spatial derivative reads as

$$\langle \nabla^\beta f^\alpha \rangle_i = \frac{\partial f_i^\alpha}{\partial x_i^\beta} = \sum_j \frac{m_j}{\rho_j} f_j^\alpha \nabla_i^\beta w(r_{ij}) = \sum_j \frac{m_j}{\rho_j} f_j^\alpha \frac{\partial w(r_{ij}, h)}{\partial x_i^\beta}, \tag{2.23}$$

where tensorial notation with Greek indices notates the spatial coordinates $\alpha, \beta = x, y, z$. The first order derivative is also second order accurate with respect to the smoothing length h under the same restrictions as discussed for the particle approximated kernel interpolation in section 2.1.2. Throughout this work, the following equivalent expressions of the directional derivative of the kernel functions are used

$$\nabla' w(\boldsymbol{x} - \boldsymbol{x}', h) = \nabla_j w_{ij} = \frac{\partial w_{ij}}{\partial x_j} = \frac{\partial w_{ij}}{\partial r_{ij}} \frac{\partial r_{ij}}{\partial x_j} = \frac{\partial w_{ij}}{\partial r_{ij}} \frac{x_j - x_i}{r_{ij}}, \tag{2.24}$$

with the norm of the distance between the two particles i and j being defined as $r_{ij} = |x_j - x_i|$.

In general, kernel approximations for higher order derivatives can be obtained following the same formalism as presented above. Subsequently, the kernel approximation of the second derivative of a field function is shown, since it plays an important role for modeling physical dissipation and fluid incompressibility, as will be seen later in section 2.7.2 and 3.1. After integration by parts, Gauss theorem and appropriate simplification, the kernel approximate for the second order derivative results in

$$\left\langle \frac{\partial^2 f(\boldsymbol{x})}{\partial \boldsymbol{x}^\alpha \partial \boldsymbol{x}^\beta} \right\rangle = - \int_\Omega f(\boldsymbol{x}') \frac{\partial^2 w(\boldsymbol{x} - \boldsymbol{x}', h)}{\partial \boldsymbol{x}^\alpha \partial \boldsymbol{x}^\beta} \, d\boldsymbol{x}'. \tag{2.25}$$

The kernel function w has to be at least continuously differentiable twice and the accuracy of the kernel estimate of the derivate is also found to be second order accurate in the smoothing length h under the assumptions of a full support domain [52, 69]. In order to obtain the particle approximated second order derivative of the field function $f(\boldsymbol{x})$, the same procedure as presented in the section 2.1.2 is applied. So, the following expression is found using the abbreviations and assumptions introduced above and dropping the brackets $\langle \rangle$ for simplicity

$$\frac{\partial^2 f_i}{\partial \boldsymbol{x}_i^\alpha \partial \boldsymbol{x}_i^\beta} = \sum_j \frac{m_j}{\rho_j} f_j \frac{\partial^2 w(r_{ij}, h)}{\partial \boldsymbol{x}_i^\alpha \partial \boldsymbol{x}_i^\beta}. \tag{2.26}$$

Using the presented formalism, consisting of kernel interpolation and particle approximation, any higher order spatial derivatives can be constructed in principle. However, the kernel function itself has to obey several conditions, which have only been touched so far. Hence, a more thorough discussion is given in the following section.

2.1.4 Kernel functions and properties

So far, the derivation of the SPH kernel interpolation and its derivatives has been presented without a detailed contemplation of the kernel function itself, even so the kernel function is the core of the interpolation algorithm. As stated by Monaghan, accuracy, stability and computational effort of the SPH method depends strongly on the chosen kernel function, which made its improvement a key topic at the beginning of the SPH development [70]. In the following, the kernel properties and requirements are reviewed.

In the first SPH paper, Gingold and Monaghan used a Gaussian kernel of the form

$$w(r_{ij}, h) = w_d e^{-(r_{ij}/h)^2} \tag{2.27}$$

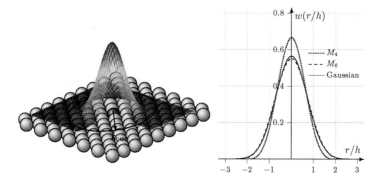

Figure 2.2: Left: Neighboring particles contributing to the kernel interpolation with compact support and cutoff radius r_{cut}. Right: M_4 and M_6 Schoenberg kernels in comparison with the Gaussian kernel function.

with the normalization constant w_d being dependent on the dimension of the computational domain. The normalization constant w_d is evaluated by

$$w_d \int_\Omega w\left(\frac{r_{ij}}{h}\right) dV = 1 \tag{2.28}$$

with $dV = [d\,\frac{r_{ij}}{h}, 2\pi\frac{r_{ij}}{h}d\,\frac{r_{ij}}{h}, 4\pi\frac{r_{ij}}{h}^2 d\,\frac{r_{ij}}{h}]$ in the one-, two- and three-dimensional case. The Gaussian kernel proved to be very accurate and stable, especially for disordered particle configurations and domains with negligible boundary effects [71]. However, since the interpolation domain is not truncated and spans over the whole spatial domain, the computational cost scales with N_p^2 with N_p being the number of particles [54]. In contrast, Lucy used a bell-shaped smoothing kernel with compact support in his first work, reducing the computational costs to $O(N \times N_p)$ with N being the number of contributing neighbors [53]. Nowadays, the use of kernels which resemble the Gaussian kernel in shape but with a compact support have become widely accepted [50].

As briefly sketched in section 2.1.1, the kernel function has to obey two conditions for the convergence of the integral estimate, which are repeated in the following.

- The normalization condition of the kernel has to be met over the support domain to assure an exact interpolation and zeroth-order consistency (i.e. exact estimation of a constant function)

$$\int_\Omega w(\boldsymbol{x} - \boldsymbol{x}', h)d\,\boldsymbol{x}' = 1. \tag{2.29}$$

- For a consistent interpolation $\langle f(x)\rangle = f(x)$, the smoothing function $w(\boldsymbol{x} - \boldsymbol{x}', h)$ has to tend towards the delta function with the smoothing length h going to zero

$$\lim_{h\to 0} w(\boldsymbol{x} - \boldsymbol{x}', h) = \delta(\boldsymbol{x} - \boldsymbol{x}'). \tag{2.30}$$

While the conditions listed above are essential for the convergence of the interpolation, the following conditions need to be obeyed for increased performance, accuracy and stability.

- As introduced by Lucy, the kernel function support should be contained as shown in figure 2.1, resulting in sparse matrices and reduced computational costs [53]

$$w(\boldsymbol{x} - \boldsymbol{x}', h) = 0 \quad \text{for } |\boldsymbol{x} - \boldsymbol{x}'| > \alpha\,h, \tag{2.31}$$

with α being a constant related to the number of piecewise splines used for the construction of the kernel as it will be shown below.

- The smoothing function should not be negative on the entire domain to exclude the occurrence of unrealistic physical quantities, e.g. a non-negative density [52].

- To achieve a good approximation of a function and its derivative, the kernel should be adequately smooth and continuous [52, 55]. Therefore, the kernel must be at least differentiable once and the derivative needs also to be continuous. Smoother functions are less sensitive to particle disorder, thereby increasing stability and accuracy [44, 49, 59].

- The smoothing function should be an even function with respect to particle distance to facilitate the conservation of momentum

$$w(\boldsymbol{x} - \boldsymbol{x}', h) = w(\boldsymbol{x}' - \boldsymbol{x}, h) = w(|\boldsymbol{x} - \boldsymbol{x}'|, h). \tag{2.32}$$

However, that specification is not satisfied if using kernel renormalization techniques as presented later in section 2.5.

- The kernel should be a strongly peaked function to adequately weight local effects.

- Particle interaction should diminish with increasing particle distance. Based on that, the kernel should be monotonically decreasing.

Due to these suggestions, the established kernels are mostly based on Schoenberg M_n splines [72], which satisfy all the above requirements and are defined by the following Fourier Transform

$$M_n = \frac{1}{2\pi} \int_{-\infty}^{\infty} \frac{\sin(0.5\,\alpha\,h)}{0.5\,\alpha\,h} \cos(\alpha\,k)\,d\alpha. \tag{2.33}$$

Transformations of several M_n splines, which are also called B-splines, into the algebraic forms are given by Schoenberg [72], as well as Monaghan [63]. Even though the second order derivative of piecewise splines can be less smooth and more complex in comparison to a Gaussian kernel, the M_4 kernel is the most widely used smoothing function in the literature. Being a piecewise cubic polynomial, it is also called cubic spline kernel, having the following form

$$w(r_{ij}, h) = w_d \begin{cases} \frac{2}{3} - \left(\frac{r_{ij}}{h}\right)^2 + \frac{1}{2}\left(\frac{r_{ij}}{h}\right)^3 & 0 \le \frac{r_{ij}}{h} \le 1 \\ \frac{1}{6}(2 - \frac{r_{ij}}{h})^3 & 1 < \frac{r_{ij}}{h} \le 2 \\ 0 & \frac{r_{ij}}{h} > 2 \end{cases} \tag{2.34}$$

where w_d is a normalization constant with $w_d = [\frac{1}{h}, \frac{15}{7\pi h^2}, \frac{3}{2\pi h^3}]$ in one, two and three dimensions, respectively [73]. The M_4 kernel resembles the Gaussian function in shape, but is more packed and peaked as shown in figure 2.2. In figure 2.3 the first and second order derivatives of the kernel function are drawn over the particle distance. It can be seen, that the first order derivative of the M_4 kernel is sufficiently smooth, but the second derivative of the kernel reduces to a piece-wise linear function. Thus, the M_4 kernel is unsuited for the direct computation of the second order derivative of a field function as presented in equation (2.26).

To obtain smooth second order derivatives, higher order spline functions are used together with an increased extent of the support [74]. The higher order polynomials have the advantage of smoother derivatives and, together with the larger extent of the support, lead to more stable simulations. This is especially true for irregular particle distributions as observed by Swegle, Morris et al. and Price [74–76]. In the presented work, the quintic and septic spline kernel have been evaluated for the evaluation of the second order derivative and are for this reason depicted in the following. The quintic

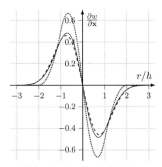
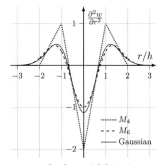

first order directional derivative   second order partial derivative

Figure 2.3: Derivatives of the cubic spline (solid line) and quintic spline kernel (dashed line).

spline or M_6 kernel takes the following form

$$
w(r_{ij}, h) = w_d \begin{cases}
(3 - \frac{r_{ij}}{h})^5 - 6(2 - \frac{r_{ij}}{h})^5 + 15(1 - \frac{r_{ij}}{h})^5 & 0 \le \frac{r_{ij}}{h} \le 1 \\
(3 - \frac{r_{ij}}{h})^5 - 6(2 - \frac{r_{ij}}{h})^5 & 1 < \frac{r_{ij}}{h} \le 2 \\
(3 - \frac{r_{ij}}{h})^5 & 2 < \frac{r_{ij}}{h} \le 3 \\
0 & \frac{r_{ij}}{h} > 3
\end{cases}
\tag{2.35}
$$

with the normalization constant w_d depending on the spatial dimension of the domain as $w_d = [\frac{1}{24}, \frac{7}{478\,\pi\,h^2}, \frac{1}{20\,\pi\,h^3}]$.

The septic spline function (M_8 kernel) is constructed out of four piecewise polynomials and is written as

$$
w(r_{ij}, h) = w_d \begin{cases}
(4 - \frac{r_{ij}}{h})^7 - 8(3 - \frac{r_{ij}}{h})^7 + 28(2 - \frac{r_{ij}}{h})^7 - 56(1 - \frac{r_{ij}}{h})^7 & 0 \le \frac{r_{ij}}{h} \le 1 \\
(4 - \frac{r_{ij}}{h})^7 - 8(3 - \frac{r_{ij}}{h})^7 + 28(2 - \frac{r_{ij}}{h})^7 & 1 < \frac{r_{ij}}{h} \le 2 \\
(4 - \frac{r_{ij}}{h})^7 - 8(3 - \frac{r_{ij}}{h})^7 & 2 < \frac{r_{ij}}{h} \le 3 \\
(4 - \frac{r_{ij}}{h})^7 & 3 < \frac{r_{ij}}{h} \le 4 \\
0 & \frac{r_{ij}}{h} > 4
\end{cases}
\tag{2.36}
$$

where the normalization factor has been numerically evaluated only for the two-dimensional case to $w_d = \frac{1}{3304\,\pi\,h^2}$ in the present work.

Besides the presented M_n Schoenberg splines, alternative kernel functions have been studied in the literature for their use in SPH simulations [71, 76, 77], but no significantly better results have been obtained in comparison to the cubic spline and quintic spline kernel as stated by Monaghan [20]. In general, it is reported, that numerical instabilities attributed to the interpolation can be effectively reduced by using higher order splines, since the stability properties strongly depend on the second derivative of the kernel [22, 78]. One also has to stress, that with an increasing order of the kernel function, the number of particles involved in the interpolation has to be increased. As a rule of thumb, the smoothing length of the kernel interpolation should be at least $h = 1.05 L_0$, with L_0 being the initial particle spacing, if an almost regular particle arrangement is maintained during the course of the simulation. Therewith, for a one-dimensional domain, four neighboring particles enter the interpolation using the cubic spline kernel, while at least ten particle interactions have to be considered using the septic spline kernel. Thus, the computational costs increase tremendously when using higher order kernels for three-dimensional simulations. Even though the choice of the extent of the interpolation domain is strongly dependent on the problem at hand, the following requirements should be considered: First, the smoothing length h needs to be large enough to include a large enough number of interacting neighbor particles in the interpolation. In this way, the discretization error of the Monte-Carlo integration is reduced and the negative effect of particle disorder on the approximation quality is also effectively decreased [61, 79]. A smoothing length to particle spacing ratio of $h/L_0 = 1.6$ has been identified by Price to be a good compromise between accuracy, stability and computational cost for the cubic spline kernel [76]. However, one has to bear in mind, that the accuracy of the kernel approximation can also be degenerated if a too large smoothing length is used, since local physical quantities are smoothed and smeared out unduly. Hence, the interaction radius should be smaller than the smallest representative length scale.

In this section, the commonly used kernel functions have been reviewed and the requirements for an exact SPH interpolation and stable simulation have been presented. Together with the kernel interpolation and the SPH derivatives presented in the sections above, the laws of continuum mechanics can be discretized within a meshless Lagrangian framework. Yet, the basic SPH formulation of the spatial derivatives presented in section 2.1.3 is inconsistent, since equation (2.23) and (2.26) cannot reproduce constant or higher order functions exactly. Thus, various SPH formulations for spatial derivatives have been developed in the literature and the different formulations will be presented and discussed with regard to their application for morphogenesis simulations in the subsequent section.

2.2 Improved SPH derivatives for continuum mechanics

In the previous section, the discretization of spatial derivatives without the use of a grid-topology, solely based on interacting particles, has been presented. However, the derived expressions do not obey the conservation principles of continuum mechanics, nor do they reproduce constant or higher order functions. Therefore, several modifications to the basic SPH derivatives based on the anti-symmetric nature of the first order derivative of the kernel function have been proposed in the literature [63]. In the following section, the practically used differential operators are summarized, together with their associated advantages and disadvantages. The approximations are verified using their ability to replicate specified functions, the so called reproducing condition. Due to its simple evaluation, the reproducing conditions are often used to classify SPH derivatives [80].

2.2.1 First order derivatives

By using the formalism described in section 2.1.3, it is possible to derive exact derivatives of an approximated function based on the function value itself and the derivative of the smoothing kernel. Even though the first order derivative shown in equation (2.23) can be derived in a consistent way from the kernel interpolation, the expression does not adhere the conservation convention, nor does it vanish for a constant field function f. In order to assure e.g. a divergence free flow field for a uniform velocity distribution, Monaghan introduced a differentiable function Φ [63]. And after applying the product rule for the term $\Phi \cdot f$, the following expression emerges [20]

$$\nabla \cdot f(\boldsymbol{x}) = \frac{1}{\Phi}(\nabla \cdot (\Phi f(\boldsymbol{x})) - f \cdot \nabla\Phi). \tag{2.37}$$

Now, the conventional SPH formalism is exerted on the modified equation, which results in a new expression for the gradient operator for particle i

$$\langle \nabla \cdot f \rangle_i = \frac{1}{\Phi_i} \sum_j \frac{m_j}{\varrho_j} \Phi_j (f_j - f_i) \cdot \nabla_i w(r_{ij}, h). \tag{2.38}$$

By choosing an appropriate term for Φ, two commonly used SPH operators for the first derivative are obtained. For $\Phi = 1$, the latter expression results in the following first order spatial derivative

$$\langle \nabla \cdot f \rangle_i = \sum_j \frac{m_j}{\varrho_j} (f_j - f_i) \cdot \nabla_i w(r_{ij}, h), \tag{2.39}$$

while choosing $\Phi = \varrho$ yields the expression

$$\langle \nabla \cdot f \rangle_i = \frac{1}{\varrho_i} \sum_j m_j (f_j - f_i) \cdot \nabla_i w(r_{ij}, h). \tag{2.40}$$

Both formulas are possible SPH discretization for the first order spatial derivative and are zeroth order and first order consistent, since the gradient vanishes for a constant, respectively linear function f. However, equation (2.39) involves the density of neighboring particles ϱ_j in the summation, while equation (2.40) is only dependent on the density of the regarded particle ϱ_i. According to Colagrossi and Landrini, equation (2.39) is more accurate and stable for multiphase systems with large density ratios [81]. Yet, both terms do not inherently obey the conservation principle, as exemplary shown for equation (2.39) by comparing the force particle i exerts on particle j F_{ij}

$$\boldsymbol{F}_{ij} = m_i \frac{d\boldsymbol{v}_i}{dt} = \frac{m_i m_j}{\varrho_j} (f_j - f_i) \frac{\partial w(r_{ij}, h)}{\partial \boldsymbol{x}_i} \tag{2.41}$$

and vice versa

$$\boldsymbol{F}_{ji} = m_j \frac{d\boldsymbol{v}_j}{dt} = \frac{m_j m_i}{\varrho_i} (f_i - f_j) \frac{\partial w(r_{ij}, h)}{\partial \boldsymbol{x}_j}. \tag{2.42}$$

With making use of the antisymmetry of the kernel function $\nabla_i w(r_{ij}, h) = -\nabla_j w(r_{ji}, h)$, one can see, that $\boldsymbol{F}_{ij} \overset{!}{=} -\boldsymbol{F}_{ji}$ is only fulfilled for $\varrho_j = \varrho_i$.

In order to ensure the conversation principle, further formulations of the first order derivative have been developed in the literature by applying the SPH discretization procedure to the identity [49]

$$\frac{\nabla \cdot f}{\varrho} = \nabla \cdot \left(\frac{f}{\varrho} \right) + \frac{f}{\varrho^2} \cdot \nabla \varrho. \tag{2.43}$$

This results in the expression

$$\langle \frac{\nabla \cdot f}{\varrho} \rangle_i = \sum_j \frac{m_j}{\varrho_j^2} f_j \cdot \nabla_i w(r_{ij}, h) + \frac{f_i}{\varrho_i^2} \sum_j \frac{m_j}{\varrho_j} \varrho_j \cdot \nabla_i w(r_{ij}, h)$$
$$= \sum_j m_j \left(\frac{f_j}{\varrho_j^2} + \frac{f_i}{\varrho_i^2} \right) \cdot \nabla_i w(r_{ij}, h). \tag{2.44}$$

By comparing the force particle i exerts on particle j, Newtons third axiom, $\boldsymbol{F}_{ij} \overset{!}{=} -\boldsymbol{F}_{ji}$ and for this reason the conservation principle is inherently satisfied due to the pairwise symmetry of the formulation. However, if using equation (2.44), the gradient of a constant field function does not inherently vanish. This is the case at free surfaces for example, where a finite numerical value is retained even for a constant function f due to the lack of symmetry in the particle configuration. The error is thus controlled by the deviation of the following equation from zero

$$\sum_j m_j \left(\frac{1}{\varrho_i^2} + \frac{1}{\varrho_j^2} \right) \cdot \nabla_i w(r_{ij}, h) \approx 0. \tag{2.45}$$

Furthermore, Monaghan introduced a generalized form of the identity by including the parameter λ [49]

$$\frac{\nabla \cdot f}{\varrho} = \frac{1}{\varrho^{2-\lambda}} \nabla \cdot \frac{f}{\varrho^{\lambda-1}} + \frac{f}{\varrho^\lambda} \cdot \nabla \left(\frac{1}{\varrho^{1-\lambda}} \right). \tag{2.46}$$

For a chosen value of $\lambda = 2$, equation (2.44) as shown above is obtained. With $\lambda = 1$, an additional expression for the first order derivative

$$\langle \nabla \cdot f \rangle_i = \sum_j \frac{m_j}{\varrho_j} (f_j + f_i) \cdot \nabla_i w(r_{ij}, h) \tag{2.47}$$

is found, which also obeys the conservation principle. Since the latter equation depends only on the particle volumes, rather than particle masses and density, the numerical noise in regions with large density differences is effectively reduced [82]. However, both equations, equation (2.44) and (2.47), are not zeroth order consistent, i.e. constant functions are not exactly estimated.

For sake of completeness, one has to mention, that further first order derivative discretization approaches can be found in the literature. For example, the same formula as shown in equation (2.47) has been obtained by Flebbe et al. using a different approach [83]. Starting from the equation (2.22), he introduced an additional integration constant $\hat{f}(\boldsymbol{x})$, obtaining the following equation

$$\langle \nabla' \cdot f(\boldsymbol{x}) \rangle = - \int_\Omega (f(\boldsymbol{x}') + \hat{f}(\boldsymbol{x})) \cdot \nabla w'(\boldsymbol{x} - \boldsymbol{x}', h) d\boldsymbol{x}'. \tag{2.48}$$

In order to guarantee a symmetric structure of the gradient formula, the integration constant is chosen to be $\hat{f}(\boldsymbol{x}) = f(\boldsymbol{x})$, which results again in equation (2.47). More first order Sph derivatives can be found in the literature, e.g. constructed by choosing different values for the parameter λ introduced in equation (2.46) [84, 85]. Though, none of the available (uncorrected) Sph first order derivatives are both, first order consistent and conservative. Hence, appropriate derivatives have to be chosen depending on the problem at hand. In addition, the presented discretization schemes should not be combined arbitrarily, but matching schemes need to be derived from a Lagrangian variational principle as shown in section 2.6 [86–89]. Prior to this, the practically relevant second order derivative formulations are presented in the following paragraph.

2.2.2 Second order derivative

In section 2.1.3, the SPH formalism to derive the basic second order derivate was shown. That approach results in equation (2.26), which is repeated here as

$$\langle \frac{\partial^2 f}{\partial x^\alpha \partial x^\beta} \rangle_i = \sum_j \frac{m_j}{\varrho_j} f_j \frac{\partial^2 w(r_{ij}, h)}{\partial x_i^\alpha \partial x_i^\beta}. \tag{2.49}$$

Similar to equation (2.23), the basic second order derivative does not conserve momentum nor does it reproduce constant functions. By using the identity

$$\frac{\partial^2 f}{\partial x^\alpha \partial x^\beta} = \frac{\partial^2 f}{\partial x^\alpha \partial x^\beta} - f \frac{\partial^2 1}{\partial x^\alpha \partial x^\beta}, \tag{2.50}$$

the second order derivative can be rewritten as [69, 75]

$$\langle \frac{\partial^2 f}{\partial x^\alpha \partial x^\beta} \rangle_i = \sum_j \frac{m_j}{\varrho_j} (f_j - f_i) \frac{\partial^2 w(r_{ij}, h)}{\partial x_i^\alpha \partial x_i^\beta}. \tag{2.51}$$

Even though, the presented version of the second order derivative was used in several works for the simulation of low Reynolds flows [23, 90, 91], all versions based on the second order derivative of the kernel functions are very noisy and sensitive to particle disorder [20]. The reason is found in the immanent small noise introduced by the kernel convolution, which is reflected and amplified in its derivatives. Moreover, the second order derivative changes its sign at a certain particle spacing as shown in figure 2.3. When the inter-particle distance is falling below the root, unphysical behavior has to be expected or oscillations in the numerical approximation occur, if the particle spacing is moving around this point [20, 92].

An alternative approach for approximating second order derivatives is the nested differentiation of the field function f. For example, the first order derivative of f is calculated using equation (2.39). With the obtained gradient, the second order derivative is calculated by repeated use of equation (2.39) to

$$\nabla^2 f_i = \sum_j \frac{m_j}{\varrho_j} \left(\nabla f_j - \nabla f_i \right) \nabla_i w(r_{ij}, h), \tag{2.52}$$

with ∇f_j and ∇f_i being the SPH discretized first order derivative. This approaches was used by several authors for modeling viscous flow and heat conduction [83, 87, 88, 93–99]. However, due to the nested differentiation, respectively summation, this approach is costly in a computational sense. In addition, instabilities are reported in the literature for non-smooth functions, which are associated to the augmented domain resulting from the nested summation by Fathei *et al.* [100]. Nevertheless, the presented approach is the method of choice if simulating matter with complex material behavior, e.g. flow of viscoelastic materials [101–111], as well as for the simulation of elastic materials using the stress based formulation of the momentum equation [75, 78, 112].

A popular scheme for approximating the second order derivatives of a scalar value f was first introduced by Brookshaw [113]. Using a Taylor expansion of the field variable $f(\boldsymbol{x}')$ around position $\boldsymbol{x}$ and under the assumption of a normalized and even kernel, one yields the following discretized form for a second order derivative

$$\langle\frac{\partial^2 f}{\partial x^\alpha \partial x^\alpha}\rangle_i = 2\sum_j \frac{m_j}{\varrho_j}\frac{(f_j - f_i)}{r_{ij}^2}(x_j^\alpha - x_i^\alpha)\frac{\partial w(r_{ij},h)}{\partial x_i^\alpha}. \qquad (2.53)$$

Besides the Taylor expansion based deviation, the second order derivative can also be obtained by inserting a particle based *Finite Difference* gradient scheme (e.g. as used in the Moving Particle Semi-Implicit method [114])

$$\langle\frac{\partial f}{\partial x^\alpha}\rangle_i = \frac{(f_j - f_i)}{r_{ij}}\frac{(x_j^\alpha - x_i^\alpha)}{r_{ij}} \qquad (2.54)$$

into the symmetric SPH gradient approximation depicted in equation (2.44) [22].

Espanol *et al.* generalized that approach to vectorial quantities. Depending on the dimension d, the following equation is obtained [115],

$$\langle\frac{\partial^2 f}{\partial x^\alpha \partial x^\beta}\rangle_i = -\sum_j \frac{m_j}{\varrho_j}(d\,e_{ij}^\alpha\,e_{ij}^\beta - \delta_{\alpha\beta})\frac{(x_j^\alpha - x_i^\alpha)}{r_{ij}^2}\frac{\partial\,w(r_{ij},h)}{\partial\,x_j^\alpha}(f_j - f_i) \qquad (2.55)$$

with the introduced notation $e_{ij} = \frac{r_{ij}}{|r_{ij}|}$ and the factor d having the value $d = 4$ in two dimensions and $d = 5$ in three dimensions, respectively. Based on the first order derivative of a monotone kernel function, the presented second order derivative does not show oscillating behavior as the latter two models (equation 2.51 and 2.52) do [100]. Therefore, equation (2.55) is frequently used in the literature for modeling heat conduction and fluid viscosity [22, 79, 100, 116–123]. However, Graham and Hughes have shown, that the smoothing length for the latter formulation has to chosen slightly larger in order to achieve a similar accuracy and ensure convergence. This observation is especially true for disordered particle configurations [79]. Furthermore, it has been shown by Colagrossi and coworker, that the presented formulation can be singular at free surfaces under certain conditions [124, 125]. This has also been reported by Schwaiger *et al.* [126].

In the present section, different formulations for the first and second order derivatives within the Smoothed Particle framework have been presented together with their associated advantages and drawbacks. It was pointed out, that the most appropriate formulation has to be chosen depending on the regarded application. For example, equation (2.40) reproduces linear functions exactly and is for this reason considered as first order consistent in the context of this work. This is a clear advantage in comparison to the basic formulations presented in section 2.1. Indeed, equation (2.40) is only first order consistent under the assumption of a complete support domain. Considering the application of the SPH formalism to the simulation of open-porous materials, the support domain may be frequently truncated by a free surface or boundary in general. Based on that, the consistency of the SPH formulations with respect to truncated domains is considered in detail in the following paragraph.

2.3 Consistency of the SPH approximation at boundaries

In the following section, the accuracy of the SPH interpolation and discretization will be analyzed with respect to its application to the formation of open-porous materials with free surfaces[1]. As stated in section 2.1.1, the kernel interpolation is reported to be of second order accuracy under the presumption of an even kernel function. However, especially if considering open-porous materials, the kernel function may not be an even function due to the truncation of the support by the free surface as shown in figure 2.4. Therewith, the order of approximation is strongly reduced as it will be shown below.

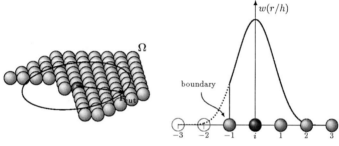

domain truncated by boundary	clipping of the constricted 2D domain

Figure 2.4: Left: Two-dimensional representation of a particle ensemble close to a boundary, e.g. free surface, together with the influence radius of a selected particle r_{cut}. Right: Clip of the two-dimensional domain and particle approximation for the particle close to the domain boundary displayed together with the course of the kernel function (M_6 kernel) [52].

To analyze the ability of the SPH discretized equations to represent the partial differential equation of the physical system, Belytschko *et al.* transfered the concept of consistency from the Finite Difference Method (known as completeness in Galerkin methods, e.g. Finite Element Method) to SPH and particle based methods in general [44, 80]. In the concept of Finite Difference discretization, a numerical interpolation scheme is consistent,

---

[1]Multiphase gas-liquid system, where the gaseous phase is not considered in general or resolved using the SPH formalism; this simplification is justified in a multiphase system, when the density ratio between the phases is large enough to consider the heavier phase as the one governing the system evolution.

if it reproduces the partial differential equations exactly in the limit of infinite grid points
and the distance between neighboring grid points going to zero [127]. And according
to the Lax equivalence theorem, the convergence of a stable and well posed numerical
model is determined by the consistency of the approximation, making the consistency of
the SPH approximation crucial for convergence [44, 52, 80, 87, 128, 129]. Though, the
consistency of a meshfree method is difficult to assess due to the irregular distribution
of particles, respectively integration nodes, in comparison to a Finite Difference scheme
on a uniform grid. Therefore, it is common to investigate the consistency of the SPH
kernel and particle approximation using the simplified concept of *reproducing condition*
instead [17, 44, 52, 80, 130]. According to Belyschko *et al.* the *reproducing condition*
is defined as the capability of the approximation to reproduce a specified function, i.e.
a polynomial [80]. Therewith, an approximation is consistent to m-th order, if any
polynomial up to order m is reproduced exactly. Based on that, the consistency of the
SPH approximation can be determined by Taylor expansion of $f(\boldsymbol{x}')$ in the vicinity of
$\boldsymbol{x}$ and inserting the result in the kernel interpolation shown in equation (2.2). In two
dimensions with $\boldsymbol{x} = (x, y)$ and under the assumption of a constant smoothing length h,
one yields

$$
\begin{aligned}
\langle f(\boldsymbol{x}) \rangle &= \int_\Omega f(\boldsymbol{x}') w(\boldsymbol{x} - \boldsymbol{x}', h) d\boldsymbol{x}' \\
&= \int_\Omega w(\boldsymbol{x} - \boldsymbol{x}', h) \sum_{k=0}^{\infty} \frac{1}{k!} \left((x' - x)\frac{\partial}{\partial x} + (y' - y)\frac{\partial}{\partial y} \right)^k f(\boldsymbol{x})\, d\boldsymbol{x}' \\
&= f(\boldsymbol{x}) \int_\Omega w(\boldsymbol{x} - \boldsymbol{x}', h)\, d\boldsymbol{x}' + ... \\
&\quad + \frac{\partial f(\boldsymbol{x})}{\partial x} \int_\Omega (x' - x) w(\boldsymbol{x} - \boldsymbol{x}', h)\, d\boldsymbol{x}' + ... \\
&\quad + \frac{\partial f(\boldsymbol{x})}{\partial y} \int_\Omega (y' - y) w(\boldsymbol{x} - \boldsymbol{x}', h)\, d\boldsymbol{x}' + ... \\
&\quad + \frac{1}{2}\frac{\partial^2 f(\boldsymbol{x})}{\partial x^2} \int_\Omega (x' - x)^2 w(\boldsymbol{x} - \boldsymbol{x}', h)\, d\boldsymbol{x}' + ... \\
&\quad + \frac{\partial^2 f(\boldsymbol{x})}{\partial x \partial y} \int_\Omega (x' - x)(y' - y) w(\boldsymbol{x} - \boldsymbol{x}', h)\, d\boldsymbol{x}' + ... \\
&\quad + \frac{1}{2}\frac{\partial^2 f(\boldsymbol{x})}{\partial y^2} \int_\Omega (y' - y)^2 w(\boldsymbol{x} - \boldsymbol{x}', h)\, d\boldsymbol{x}' \\
&\quad\ \vdots \\
&\quad + O(h^3)
\end{aligned}
\tag{2.56}
$$

The integral terms in equation (2.56) are called moments of the interpolation function. For an interpolation of order m, the kernel function, respectively its moments have to satisfy

$$\int_\Omega w(\boldsymbol{x} - \boldsymbol{x}', h) d\boldsymbol{x}' = 1$$
$$\int_\Omega (x' - x) w(\boldsymbol{x} - \boldsymbol{x}', h) d\boldsymbol{x}' = 0$$
$$\int_\Omega (y' - y) w(\boldsymbol{x} - \boldsymbol{x}', h) d\boldsymbol{x}' = 0$$
$$\int_\Omega \frac{1}{2}(x' - x)^2 w(\boldsymbol{x} - \boldsymbol{x}', h) d\boldsymbol{x}' = 0 \qquad (2.57)$$
$$\vdots$$
$$\int_\Omega \frac{1}{m!}(y' - y)^m w(\boldsymbol{x} - \boldsymbol{x}', h) d\boldsymbol{x}' = 0$$

The zeroth moment, resembled by the first equation in (2.57), is satisfied by the normalization condition of the kernel as stated in section 2.1.1 and 2.1.4 for a complete integration domain. The two first order moments (second and third equation in (2.57)) are complied by an even kernel function, again under the presumption of a complete integration domain. However, due to the requirement of a positive kernel function as stated in section 2.1.4, equation four and following cannot be fulfilled and the interpolation is limited to second order accuracy or first order consistency. This results in the exact reproduction of constant (zeroth order consistency) as well as linear functions (first order consistency) by the kernel approximation.

Thus, constant and linear functions are only reproduced exactly for a complete support domain, e.g. for bulk particles. If the domain is truncated by a boundary as shown in figure 2.4, the kernel approximation is not even zeroth order consistent, since the normalization condition is not fulfilled. One has to stress, that this is a serious drawback if dealing with open-porous materials and thin structures in general, as the numerical scheme is not consistent in these prominent regions and the correctness of the overall simulation results are in doubt.

As mentioned in section 2.1.2, the accuracy of the Sph formalism is also influenced by the particle approximation. To ensure the maximum possible first order consistency, the discrete equation of the normalization condition,

$$\sum_j \frac{m_j}{\varrho_j} w(r_{ij}, h) = 1 \tag{2.58}$$

and the discrete equation of first order consistency condition

$$\sum_j \frac{m_j}{\varrho_j} (\boldsymbol{x} - \boldsymbol{x}_j) w(r_{ij}, h) = 0 \tag{2.59}$$

have to be fulfilled. However, both conditions are also violated at support domains truncated by boundaries as exemplary shown in figure 2.4 on the right, due to the inconsistency introduced by the kernel interpolation. In addition, an irregular distribution of particles can lead to the violation of both consistency conditions even in interior regions of the domain as discussed in detail in section 2.1.2. Yet, it has been shown by Quinlan *et al.*, that the first order consistency is retained for irregularly distributed bulk particles, whose spatial perturbations from the regular and equispaced configuration are small with respect to the smoothing length. And in the present study, a relatively regular and equispaced particle distribution is maintained throughout the simulation due to the expected low particle dynamics in the desired application. Moreover, an incompressible SPH approach is used, which prevents particle clustering and aids to conserve an equispaced particle setup. Therewith, the assumption of second order accuracy for bulk particles is justifiable in the present application. Hence, no further discussion of particle disorder and its influence of the accuracy of the SPH interpolation is made and the interested reader is referred to [20, 61, 63–65, 68, 91].

To conclude, the SPH formalism is of second order accuracy or first order consistency regarding a regular as well as moderately irregular particle distribution in the interior of the domain. For strongly irregular particle distributions, which are not expected in the presented work, the SPH approximation is not consistent. Furthermore, the SPH approximation is not consistent in regions, where the interpolation domain is truncated by boundaries. This is a crucial fact, since the formation of open-porous structures is governed by evolution of interfaces as well as free surfaces. Therefore, special care is given in this work to the description of multiphase systems and free surfaces in particular. The utilized approaches in this work can be classified in two categories. In the first category, the standard SPH formalism, as presented above, is used and the error introduced by the truncated domain is attenuated by the use of special boundary treatment strategies. A detailed description of the available approaches is given in the following section 2.4. In the second category, the consistency of the numerical method is restored in general by modification the kernel function using so called *Corrected SPH* approaches. A review of the available corrections will be presented in section 2.5, with a focus on the description of the methods used in the presented work.

2.4 Boundary treatment within the SPH framework

In general, the treatment of boundaries is a weak point within the SPH formalism. As stated above, an exact SPH interpolation is only achieved in regions with full particle support. In contrast, if the interpolation domain is truncated by a boundary, as shown in figure 2.5, the interpolation is inaccurate, driving the need for special boundary treatment strategies, which will be sketched below.

In boundary near regions, the kernel approximation shown in equation (2.15) is divided into the classical SPH part and a term reconstructing the missing boundary

$$f_i = \sum_j \frac{m_j}{\varrho_j} f_j w_{ij} + \int_{\Omega_b} f_b(\boldsymbol{x}') w(\boldsymbol{x} - \boldsymbol{x}') d\,\boldsymbol{x}', \qquad (2.60)$$

with f_b being the continuation of the field variable f behind the boundary and Ω_b the volume of the boundary region as depicted in figure 2.5. As presented in section 2.1.3, a boundary integral also emerges during the derivation of the first order derivative as shown in equation (2.20). The spatial derivative is then transfered to

$$\begin{aligned} \langle \nabla f \rangle_i &= -\sum_j \frac{m_j}{\varrho_j} f_j \nabla_j w_{ij} - \int_{\Omega_b} f_b(\boldsymbol{x}') \nabla' w(\boldsymbol{x} - \boldsymbol{x}') d\,\boldsymbol{x}' \\ &= -\sum_j \frac{m_j}{\varrho_j} f_j \nabla_j w_{ij} + \int_S f_b(\boldsymbol{x}')\, w(\boldsymbol{x} - \boldsymbol{x}', h) \cdot \hat{n}\, d\,\boldsymbol{S}', \end{aligned} \qquad (2.61)$$

with the influence of the boundary appearing in the second term. Therewith, the boundary treatment approach needs to reconstruct the interpolation domain on the one hand, but on the other hand, it needs to enforce the mathematical boundary condition in addition. In general, this is seen as one of the main drawbacks of meshless methods in comparison to grid-based methods, since the emerging problems at the boundary are only rudimentary solved and no general and cost-effective methodology is available so far. In general, the most commonly used approaches in the literature can be classified into three categories.

In the first category, the arising boundary terms are approximated by analytical expressions [131, 132] or by extrapolation of the regarded field variable across the boundary as done in the *normal flux method* [133, 134]. At least for planar boundaries, relatively simple analytical solutions can be found for the boundary terms of equation (2.60) and (2.61), which are only dependent on the distance of the regarded particle to the

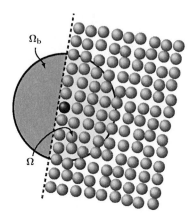

Figure 2.5: Divided interpolation domain truncated by a boundary. Part of the interpolation domain coincide with the simulation domain Ω, while the other part of the interpolation domain lies exterior to the boundary Ω_b.

boundary. In principle, the analytical boundary term can represent solid boundaries as well as free surface boundaries. However, the approach is limited to rather simple geometries and for this reason not applicable for the problem at hand. A detailed description of this approach and analytical expression are given in [61, 131]. In contrast, the *normal flux method* developed by De Leffe *et al.* can be applied to arbitrarily shaped solid boundaries. But, the kind of predominated forces acting on the solid boundary, e.g. the body force, has to be known a priori in order to be able to extrapolate the value of the force across the boundary [133].

For modeling solid wall boundaries, repellent boundary particles are commonly used [135]. In doing so, moving particles close to a solid surface experience an external repulsive force, which is dependent on the distance to the obstacle. Frequently, a Lennard-Jones type potential is applied, which has to be parameterized depending on the problem at hand [20, 136, 137]. Though, the method is known of producing unphysical numerical noise in the pressure and stress field [134].

For modeling evolving boundaries, whose positions are not known a priori, a dynamic reconstruction of the boundary has to be performed and the *virtual particle* or *ghost*

free surface particles   virtual particles

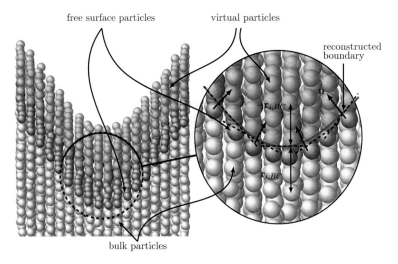

reconstructed
boundary

bulk particles

Figure 2.6: Arrangement of virtual particles ⬤ at a free surface. The virtual particle layer has to be at least as deep as the cutoff radius of the kernel interpolation. The virtual particles ⬤ are constructed by reflection of the bulk ⬤ and boundary particles ⬤ at the reconstructed free surface boundary. The reflection of a bulk particle at the reconstructed boundary is shown in the detail on the right. The boundary is reconstructed based on the orthogonal to the surface normal vectors of the free surface boundary particles.

particle approach is the method of choice [22, 23, 61, 93, 131, 138]. If no explicit physical boundary positions are given, the distribution of the virtual particles has to be computed from the distribution of the particles inside the simulation domain. This can be done by reflecting any particle, whose interpolation domain is truncated by the boundary, at the interface tangent line to the boundary. To exclude artificial boundary effects on the interior particles, the virtual particle boundary layer has to be as thick as the cutoff radius of smoothing function. Depending on whether reflective boundaries, e.g. solid walls, or transmissive boundaries, e.g. free surfaces, are considered, the respective property of the virtual particle has to be adjusted. For free surfaces, the regarded property of the virtual particle, e.g. velocity, is identical to the one of the original particle. For reflective boundaries, the normal component of the property has to change sign [138]. One possible, rigorous approach of generating the virtual particle distribution is the so called *multiple boundary tangent* method developed by Yildiz *et al.* and refined by Shadloo *et al.* [139, 140]. The main principle of this approach is depicted in figure 2.6. After identifying the boundary, the tangent line to the boundary is calculated from the unit normal vector for every particle on that boundary (boundary particle). The neighboring fluid particles of the regarded boundary particle are mirrored with respect to the corresponding boundary tangent. However, if a fluid particle, which needs to be reflected, is closer to another boundary particle, the fluid particle gets reflected at the boundary tangent line of the closest boundary particle. Therewith, the virtual particle distribution is generated using multiple tangent lines. Subsequently, boundary near particles (particles, which are not on the boundary, but whose interpolation domain is truncated by the boundary) are associated to the closest boundary particles, inheriting the virtual particle neighbors of the respective boundary particle. By using this approach, which is developed for complex solid wall boundaries but can be transfered to free boundaries in principle, the particle deficit at the boundary is eliminated and the accuracy of the kernel interpolation is restored. But, the approach is cumbersome to implement in a numerical code and worse, features enormous numerical costs with regard to computation time and storage. This is especially true for its application to systems dominated by boundaries. Nonetheless, the approach can be simplified for simple geometries by reflecting all neighboring fluid particles at the respective boundary tangent line, as done by Morris for assuring zero velocity slip at solid boundaries [22]. Nevertheless, the computational costs for restoring the particle deficit using virtual particles are comparably high for applications dominated by free surfaces. Therefore, an approach for restoring particle consistency in general, using so called *Corrected* SPH approaches, is shown in the following section.

2.5 Corrected SPH approaches

As stated in the previous section, the Smoothed Particle Hydrodynamics method is not consistent in regions where the support domain is constricted by boundaries. However, for the simulation of open-porous materials, with possibly thin structures and ligaments, regions where the interpolation domain is truncated by a free surface, can be large or even dominant. To eliminate the particle deficiency associated with the free surface, a corrected SPH approach, called Corrective Smoothed Particle Method (CSPM) [141–144] is used in conjunction with a renormalized smoothing kernel in the present work. The CSP method was introduced by Chen and coworkers and combines the SPH typical kernel estimate with a Taylor series expansion, thereby restoring the consistency of the particle method, while moderately increasing the computational costs.

One has to stress, that there are several corrections to SPH or advancements of SPH available in the literature, which can be classified into two categories. In the first group, the smoothing function is reconstructed for each particle to satisfy the consistency conditions. Examples thereof are the Element Free Galerkin Method [145], Reproducing Kernel Particle Method [146], Moving-Least-Square-Particle Hydrodynamics [147, 148], Meshless Local Petrov Galerkin Methods [149] and Point Interpolation Method [150] to mention a few. But because of the reconstruction of the smoothing function, the resulting function can be partially negative and the methods are thus not preferred for fluid dynamic applications [44]. Approaches falling in the second category restore consistency without changing the conventional kernel function. Common representatives are for example, Normalized SPH [151], Normalized Corrected SPH [112] and Modified SPH [152] or Finite Particle Method [153] as well as CSPM. CSPM is a straightforward supplement of SPH and can be mixed with the standard scheme if needed. The concept of the correction scheme is applied and refined by renormalization of the smoothing kernel in the context of this work, and therefore presented in the following section.

As shown in section 2.3, the SPH formalism is inconsistent due to an incomplete kernel interpolation close to or on boundaries. To alleviate the particle deficiency at boundaries, Chen and Beraun combined the SPH kernel estimate concept with a Taylor series expansion [141]. By expanding a Taylor series for $f(\boldsymbol{x}')$ around a spacial point $\boldsymbol{x}$

and multiplying both sides with the kernel function $w(\boldsymbol{x} - \boldsymbol{x}', h)$, the following expression is obtained after integration over the domain Ω

$$
\begin{aligned}
\int_\Omega f(\boldsymbol{x}') w(\boldsymbol{x} - \boldsymbol{x}', h) d\,\boldsymbol{x}' = {} & f(\boldsymbol{x}) \int_\Omega w(\boldsymbol{x} - \boldsymbol{x}', h) d\,\boldsymbol{x}' \\
& + f_\alpha(\boldsymbol{x}) \int_\Omega (x'^\alpha - x^\alpha) w(\boldsymbol{x} - \boldsymbol{x}', h) d\,\boldsymbol{x}' \\
& + \frac{1}{2!} f_{\alpha\beta}(\boldsymbol{x}) \int_\Omega (x'^\alpha - x^\alpha)(x'^\beta - x^\beta) w(\boldsymbol{x} - \boldsymbol{x}', h) d\,\boldsymbol{x}' \\
& + \ldots
\end{aligned}
\tag{2.62}
$$

with $f_\alpha(\boldsymbol{x})$ being the short-cut for the spacial derivative $\frac{\partial f(\boldsymbol{x})}{\partial x^\alpha}$ and $f_{\alpha\beta}(\boldsymbol{x})$ for $\frac{\partial^2 f(\boldsymbol{x})}{\partial x^\alpha \partial x^\beta}$, while the indices α and β represent spacial components $\alpha, \beta = x, y, z$. The presented equation (2.62) is the heart of the CSPM for constructing corrected kernel estimates. To obtain the corrective kernel estimate of $f(\boldsymbol{x})$, the derivative terms in equation (2.62) are neglected and the equation is solved for $f(\boldsymbol{x})$. Therewith, the corrective kernel estimate is expressed as

$$
\langle f(\boldsymbol{x}) \rangle = \frac{\int_\Omega f(\boldsymbol{x}') w(\boldsymbol{x} - \boldsymbol{x}', h) d\,\boldsymbol{x}'}{\int_\Omega w(\boldsymbol{x} - \boldsymbol{x}', h) d\,\boldsymbol{x}'}.
\tag{2.63}
$$

After discretization using the numerical quadrature as shown in section 2.1.2, the particle based kernel approximation of a field function f for particle i is written as

$$
\langle f \rangle_i = \frac{\sum_j \frac{m_j}{\varrho_j} f_j w(r_{ij}, h)}{\sum_j \frac{m_j}{\varrho_j} w(r_{ij}, h)}.
\tag{2.64}
$$

For ordered particles lying in the interior domain, the denominator is equal to one. For this reason, equation (2.64) reduces to the conventional kernel estimate shown in equation (2.15). Since the kernel is an even function, the second term on the right hand side of equation (2.62) is equal to zero. So, the CSPM kernel approximation is second order accurate in the smoothing length h, just as the standard SPH kernel approximation. For particles on or near boundaries, the second term in equation (2.62) is not equal to zero and the truncation error is increased to the order of h. Nevertheless, the corrected kernel approximation is also consistent in regions truncated by boundaries, which is not the case in the conventional SPH framework with an error in the order of h^0. Furthermore, it has been demonstrated by Chen and Beraun, that the corrected kernel estimate yields better results than the conventional approximation, shown in equation (2.15), since the denominator also corrects the effect of an irregular particle distribution during the course of the simulation [141]. With introducing the so called Shepard kernel $\hat{w}$, equation (2.64) is simplified to [80, 154]

$$\langle f \rangle_i = \sum_j \frac{m_j}{\varrho_j} f_j \hat{w}(r_{ij}, h), \quad \text{with} \quad \hat{w}(r_{ij}, h) = \frac{w(r_{ij}, h)}{\sum_k \frac{m_k}{\varrho_k} w(r_{ik}, h)}. \tag{2.65}$$

For sake of completeness, one has to stress, that the form of the CSPM kernel approximation shown in equation (2.64) is equal to the kernel approximation of the density field for boundary particles derived in the work of Randles and Libersky [155]. The latter approach is based on a renormalization using the unity condition of the kernel estimate. Following that work, Vignjevic and coworkers derived the Normalized Corrected SPH approach [112]. A normalized kernel function was also used by Johnson *et al.* [151, 156]. Moreover, a similar kernel correction is used in the Moving-Least-Square Kernel Galerkin Method [157, 158] and applied to SPH by Bonet and Lok [87].

In contrast to the approaches of Randles and Libersky [155], Bonet and Lok [87] as well as Vignjevic *et al.* [112], the CSPM approach has the ability to educe corrected higher order derivatives. By replacing the kernel function $w(\boldsymbol{x} - \boldsymbol{x}', h)$ in equation (2.62) by its first order derivative $\nabla' w(\boldsymbol{x} - \boldsymbol{x}', h)$ and neglecting the second order derivative terms, one yields the integral estimate of $\nabla' f(\boldsymbol{x})$. In the discretized particle syntax, the first order derivative takes the following form with Einstein summation implied

$$\frac{\partial f_i}{\partial \boldsymbol{x}_i^\alpha} \sum_j \frac{m_j}{\varrho_j} (\boldsymbol{x}_j^\alpha - \boldsymbol{x}_i^\alpha) \frac{\partial w(r_{ij}, h)}{\partial \boldsymbol{x}_j^\beta} \approx \sum_j \frac{m_j}{\varrho_j} (f_j - f_i) \frac{\partial w(r_{ij}, h)}{\partial \boldsymbol{x}_j^\beta}. \tag{2.66}$$

Equation (2.66) represents d coupled equations, with d being the spatial dimension of the problem. To obtain the corrective approximation of the first order derivative of particle i, the following system of equations has to be solved simultaneously [143]

$$\boldsymbol{A}_{\alpha\beta,i}\, \boldsymbol{f}_{\beta,i} = \boldsymbol{b}_{\alpha,i}, \tag{2.67}$$

with $\boldsymbol{A}$ being a matrix with rank equal to d. $\boldsymbol{A}$ is calculated as follows

$$\boldsymbol{A} = -\sum_j \frac{m_j}{\varrho_j} (\boldsymbol{x}_j - \boldsymbol{x}_i) \otimes \nabla_i w(r_{ij}, h), \tag{2.68}$$

where $\otimes$ denotes the dyadic product. The right hand side vector $\boldsymbol{b}$ has the length dimension and the following components

$$\boldsymbol{b}_{\alpha,i} = \sum_j \frac{m_j}{\varrho_j} (f_j - f_i) \frac{\partial w(r_{ij}, h)}{\partial \boldsymbol{x}_j^\alpha}. \tag{2.69}$$

Due to the small size of the problem, the matrix inversion can be done analytically, resulting in reduced computational costs.

The accuracy of the corrected gradient operator $\frac{\partial f(x_i)}{\partial x^\alpha}$ is second order accurate in the smoothing length h for interior particles. For particles, whose interpolation domain is truncated by a boundary, the terms $\sum_j \frac{m_j}{\varrho_j}(\boldsymbol{x}_j - \boldsymbol{x}_i)^2 \nabla_i w(r_{ij}, h)$ do not vanish and therefore, the accuracy is only of first order in the smoothing length h. The CSPM first order derivative is complete to first order for all particles, reproducing constant and linear functions exactly. This is true even so the consistency criterion for the kernel approximation of $\sum \frac{m_j}{\varrho_j} \nabla_i w(r_{ij}, h) = 0$ is not satisfied for boundary particles. By using the difference of the function values of the interacting particles, as shown in section 2.2, the zeroth order consistency is obtained.

As stated above, the same corrected first order derivative has been published by several research groups, however with different motivations. Randles and Libersky introduced the second order tensor $\boldsymbol{A}$ to enforce the reproduction of linear stresses [155]. Bonet and Lok derived the tensor $\boldsymbol{A}$ to compel the conservation of linear and angular momentum [87], while Vignjevic *et al.* derived his Normalized Corrected SPH approach from the more general principles of homogeneity and isotropy of space [112]. In the present work, the CSPM approach is further developed by combination with a renormalized kernel (Shepard kernel). For the first order derivative, the combination leads to the same expression for the derivative as in the NCSPH approach of Vignjevic and coworkers. Yet, the advantage the CSPM approach lies in its sound mathematically basis, with which higher order derivatives can be estimated. Though, the corrected second order derivative based on the CSPM scheme yielded less accurate and stable results in comparison to other corrected approaches and is consequently not explicitly used in the present work. Nevertheless, the theoretical background as well as the implemented discretization is shown in appendix E.

In the present section, corrected SPH approximations using the so called CSP method have been shown. In contrast to the conventional SPH kernel estimates, the corrected kernel estimates are consistent on the whole interpolation domain. Therewith, a convergent numerical scheme is obtained (under the presumption of a stable discrete model) and the reconstruction of the interpolation domain at the domain boundaries as shown in section 2.4 is made obsolete. So far, various conventional as well as corrected SPH discretization formulas have been presented together with their immanent advantages and disadvantages. In the subsequent section, the adept combination of these formulas to an accurate and stable numerical scheme using the concept of variational principle is presented.

2.6 SPH discretization from Variational Principles

As shown in section 2.2 and 2.5, various SPH approximations can be derived, depending on the chosen approximations or transformations. However, it has been found by Bonet and Lok, that only certain combinations of the discretization scheme for the continuity equation and the divergence of the stress tensor in the momentum equation naturally ensures the conservation of linear and angular momenta. And this leads to more regular particle distributions and on these grounds to more stable as well as accurate simulations [87]. Yet, only slight improvements have been reported by Marri and White [85] as well as Hernquist and Katz [84]. Nevertheless, the discretization of the mathematical operators derived for the developed SPH approaches presented in section 3.2 are based on the variational principle and thus, the idea is reviewed in the following.

By using the so called variational principles approach, the stress forces can be derived from an internal energy functional, which inherently satisfies the conservation principle under the requirement, that the internal energy is invariant with respect to rigid body motions [87]. In the following, the derivation of a variational consistent divergence operator for the stress tensor depending on the here chosen representation of the density

$$\frac{D\varrho_i}{Dt} = \varrho_i \sum_j V_j (v_j - v_i)\nabla_j w_{ij}. \tag{2.70}$$

is briefly sketched. The basic theory and formalism presented in the current paragraph follows the work of Landau and Lifschitz as well as Price and Monaghan [20, 76, 89, 159]. The equations of motion for the particles can be derived from the Hamilton's principle or principle of least action. Therewith, the discretization scheme for the momentum equation can be obtained from a Lagrangian L, here for a non-dissipative and compressible fluid, which is simply the difference between the kinetic and internal energy [86, 160, 161]

$$L = \int \varrho \left(\frac{1}{2}v^2 - u(\varrho, s)\right) dV, \tag{2.71}$$

with u being the specific internal energy, dependent on the fluid density ϱ and specific entropy s. After discretization using the basic SPH kernel interpolation (equation (2.2)), the Lagrangian reads in SPH notation

$$L = \sum_j m_j \left(\frac{1}{2}v_j^2 - u_j(\varrho_j, s_j)\right). \tag{2.72}$$

Using the principle of least action, also called Hamilton's principle, the following expression can be derived using the Lagrangian as shown in appendix C

$$\int \left[\frac{D}{Dt}(m_i \boldsymbol{v}_i) + \sum_j m_j \frac{p_i}{\varrho_i^2} \frac{\delta \varrho_j}{\delta \boldsymbol{x}_i} \right] \delta \boldsymbol{x}_i \, dt = 0. \tag{2.73}$$

Based on the continuity equation shown above (equation (2.70)), the Lagrangian variation of the density ϱ_i upon virtual displacement results in [76]

$$\delta \varrho_i = \varrho_i \sum_j V_j (\delta \boldsymbol{x}_j - \delta \boldsymbol{x}_i) \nabla_j w_{ij}. \tag{2.74}$$

Inserting the above equation in equation (2.73), one obtains [55]

$$\int \left[\frac{D}{Dt}(m_i \boldsymbol{v}_i) - \sum_j m_j \frac{p_j}{\varrho_j} \sum_k V_k (\delta_{ki} - \delta_{ji}) \nabla_k w_{jk} \right] \delta \boldsymbol{x}_i \, dt = 0. \tag{2.75}$$

After integration by parts of the first term and assuming $\nabla_j w_{jk} = -\nabla_k w_{jk}$ for an even kernel function, the expression is simplified to

$$\int \left[m_i \frac{D \boldsymbol{v}_i}{Dt} - \frac{m_i}{\varrho_i} \sum_j V_j \left(p_i + p_j \right) \nabla_k w_{jk} \right] \delta \boldsymbol{x}_i \, dt = 0, \tag{2.76}$$

which results in

$$\int \left[\varrho_i \frac{D \boldsymbol{v}_i}{Dt} - \sum_j \frac{m_j}{\varrho_j} \left(p_i + p_j \right) \nabla_k w_{jk} \right] \delta \boldsymbol{x}_i \, dt = 0. \tag{2.77}$$

From the latter expression, the gradient operator for the stress tensor in the momentum balance

$$\langle \nabla \cdot f_i \rangle = \sum_j \frac{m_j}{\varrho_j} (f_j + f_i) \cdot \nabla_i w(r_{ij}, h) \tag{2.78}$$

can be excerpted. So, this gradient formula, which has already been derived in section 2.2, equation (2.47), is variational consistent with equation (2.39)

$$\langle \nabla \cdot f_i \rangle = \sum_j \frac{m_j}{\varrho_j} (f_j - f_i) \cdot \nabla_i w(r_{ij}, h), \tag{2.79}$$

used for discretizing the continuity equation. And as stated initially, the total linear and angular momenta are naturally conserved if the discretization scheme of the momentum equation is variational consistent with the one of the continuity equation.

51

For sake of completeness, one has to stress, that if the continuity equation is discretized using equation (2.40), the variational consistent divergence operator discretization scheme of the momentum balance is of the form shown in equation (2.44). The latter equation is also variational consistent, if the density is estimated by interpolation from neighboring masses as shown later in equation (2.96) [89].

In the present section, an approach for deriving variational consistent combinations of Sph discretization schemes has been presented. Therewith, the fundamentals for deriving conventional as well as corrected Sph approximations have been reviewed. In the following paragraph, the two most common numerical algorithms for solving fluid dynamic problems within the Sph framework are shown.

2.7   SPH algorithms for compressible and incompressible flow

As stated earlier, the SPH method was originally developed in the field of astrophysics for discretizing the Euler equations to simulate dynamical processes of compressible materials [53, 55]. Later on, the SPH approach was increasingly used for solving the equations of motion and conservation of mass in the field of solid and fluid dynamics [63, 162, 163]. For this purpose, the conventional compressible approach was adopted by Monaghan *et al.* for the simulation of incompressible materials by introducing a weakly compressible equation of state [49, 164]. In doing so, the hydrostatic pressure is calculated using an (artificial) algebraic thermodynamic equation of state, which is dependent on the relation of the actual particle density and the incompressible one [49, 135]. The advantage of that approach lies in the relative ease of programming and the low computational costs for evaluating the pressure term. But, due to the incompressible nature of the fluid, the time steps of the temporal integration need to be very small to, at least approximately, resolve the traveling sound waves in the material. This is especially of disadvantage in the present work, since rather long-time simulations are required for the description of structure formation processes. In addition, small density errors, occurring mostly at the boundaries of the domain, are amplified by the numerical speed of sound and result in non-physical pressure fluctuations. These fluctuations can lead to numerical instabilities, eventually stopping the computation as observed by Koshizuka *et al.* as well as Cummins and Rudman [92, 165]. Therefore, a truly incompressible Smoothed Particle Hydrodynamics approach based on a projection scheme has been developed by Cummins and Rudman [92], which has been used and partly advanced by several authors [140, 166–177]. Lee *et al.* compared both approaches and observed an increase in CPU time of a factor of 2 – 20 depending on the test case if using the explicit weakly compressible SPH in comparison to the truly incompressible SPH approach [174, 178]. In the following two sections, both approaches are reviewed.

2.7.1 Explicit SPH for compressible and weakly compressible materials

For the explicit Sph approach, used for the simulation of compressible as well as weakly compressible materials, mostly second order accurate predictor-corrector type integration schemes are adopted in the literature. Examples thereof are the modified Euler scheme or the symplectic Velocity Verlet integration as well as the Leap-frog method [67, 87, 89, 135, 164]. As stated by the name, these integration methods are two-step schemes, with the sequence of the integration steps exemplary shown in figure 2.7 for the modified Euler method. In the first step of the integration scheme introduced by Monaghan, all quantities are predicted for the instant in time t^* based on the current source terms [164]. These source terms are dependent on quantities of particle i as well as all neighboring particles N calculated in the previous time step t^n

$$v_i^* = v_i^n + f(v_j^n, p_j^n, F_i^n)\frac{\Delta t}{2} \quad j = 1, \ldots, N \tag{2.80}$$

$$x_i^* = x_i^n + f(v_i^n)\frac{\Delta t}{2}. \tag{2.81}$$

Using the explicit Sph approach, two ways for calculating the density are possible. In the first approach, the density is calculated based on the integration of the continuity equation, with the density of particle i being dependent on the velocities of the neighboring particles as sketched below

$$\varrho_i^* = f(v_j^n)\frac{\Delta t}{2} \quad j = 1, \ldots, N. \tag{2.82}$$

In the second approach, the density of particle i is calculated by summing up the kernel weighted mass of the neighboring particles m_j

$$\varrho_i^* = \sum_j m_j w(r_{ij}, h). \tag{2.83}$$

Since the former approach does not conserve the total mass, equation (2.83) is commonly used in the literature for single phase flow. Regardless of the chosen approach, the density is used for calculating the current isotropic pressure via an equation of state. As an example Tait's equation of state (also known as Batchelor's equation of state), which is commonly used for approximating incompressible materials as weakly compressible, is shown below below [56, 179]

$$p = \frac{\varrho_0 c_0^2}{\gamma}\left[\left(\frac{\varrho}{\varrho_0}\right)^\gamma - 1\right]. \tag{2.84}$$

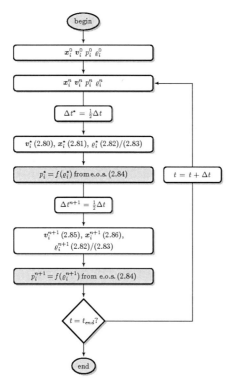

Figure 2.7: Flow sheet of the explicit SPH algorithm for compressible and weakly compressible materials.

The nominal density is abbreviated by ϱ_0 and c_0 represents the speed of sound and γ the polytropic constant. Once all variables for all particles are obtained at the intermediate time step, the source terms are evaluated at the predicted time step t^* and combined with the results of the previous time step t^n. Hence, the final corrected primitive variables at the new time step t^{n+1} are centered in time to obtain

$$v_i^{n+1} = v_i^n + f(v_j^n, v_j^*, p_j^n, p_j^*, F_i^n, F_i^*)\frac{\Delta t}{2} \quad j = 1, \ldots, N \qquad (2.85)$$

$$x_i^{n+1} = x_i^n + f(v_i^n, v_i^*)\frac{\Delta t}{2} \qquad (2.86)$$

Finally, the density and therewith pressure value at time step t^{n+1} is again estimated via equation (2.82) or (2.83). Besides these predictor-corrector style integrators, explicit forth order Runge-Kutta methods have been used mostly for simulations of multiphase systems [67, 180] and longtime astrophysical simulations [181], while implicit integration schemes are rarely used [182].

In the current paragraph, the commonly used predictor-corrector scheme for the compressible and weakly compressible SPH has been presented. In principle, a similar predictor-corrector type code structure is used for the truly incompressible SPH method, which is based on a two-step projection scheme. In the following section, the basic fundamentals of the applied projection method are summarized and the truly incompressible SPH approach is derived in a general form.

2.7.2 Incompressible SPH algorithm

In the following section, an outline of the state-of-the-art incompressible SPH approach is given. The method is based on the classical projection method for solving unsteady incompressible fluid flow. The projection method was independently introduced by Chorin as well as Temam to decouple the computation of the velocity field and the pressure field [183, 184] and is extensively used in Eulerian grid-based methods, such as Finite Difference [185, 186], Finite Volume [187] and Finite Element approaches [188]. The approach is based on the principle, that the velocity field of an incompressible fluid is divergence-free (as later shown in equation (3.3)). The projection method was first applied within the concept of SPH by Cummins and Rudman [92]. Yet, a similar procedure has been used earlier in the Moving Particle Semi-Implicit method invented by Koshizuka et al. [189]. Colin et al. also proposed a divergence free incompressible SPH algorithm, which is based on the evaluation of the second order derivative of the kernel function [190].

Using the projection method, the pressure value required to enforce the incompressibility of the fluid is obtained by projecting an intermediate velocity field onto a divergence-free velocity field [183]. The theoretical background of the projection is the so called Helmholtz-Hodge decomposition, which states that any vector field v can be split up into a divergence-free (solenoidal) part v_S and an irrotational or curl-free part v_I [191].

Curl-free vector fields, can be replaced by the gradient of the respective scalar field ϕ to give [192]

$$\boldsymbol{v} = \boldsymbol{v}_S + \boldsymbol{v}_I = \boldsymbol{v}_S + \nabla\phi \qquad (2.87)$$

By taking the divergence of equation (2.87), a Poisson equation for the scalar function ϕ is obtained, since the divergence of a solenoidal vector field vanishes. With known vector field $\boldsymbol{v}$ the Poisson equation

$$\nabla^2\phi = \nabla \cdot \boldsymbol{v} \qquad (2.88)$$

can be solved for the scalar function ϕ. The divergence-free fraction of the vector field $\boldsymbol{v}_S$ is subsequently obtained by

$$\boldsymbol{v}_S = \boldsymbol{v} - \nabla\phi. \qquad (2.89)$$

Based on Chorin's work, Cummins and Rudman applied the Helmholtz decomposition to SPH in a fractional two-step algorithm [92]. In the following, the procedure is applied to the momentum balance shown later in equation (3.5), together with a first order explicit Euler integration scheme for sake of simplicity. The respective flowchart is shown in figure 2.8. In the predictor step, the intermediate particle velocities $\boldsymbol{v}^*$ are explicitly calculated based on the momentum equation, but ignoring the pressure gradient term. The intermediate velocity is dependent on inter-particle forces, resulting from the deviatoric stress tensor $\boldsymbol{\tau}^n$ evaluated at time step t^n as well as external forces and hence is calculated as

$$\boldsymbol{v}^* = \boldsymbol{v}^n + \frac{\Delta t}{\varrho^n}\left(\nabla \cdot \boldsymbol{\tau}^n + \varrho^n\,\boldsymbol{f}^n\right). \qquad (2.90)$$

Since the intermediate velocity field $\boldsymbol{v}^*$ is independent of the pressure gradient of the previous time step t^n, the projection is called a full pressure projection. An incremental projection in contrast considers the pressure gradient of the previous time step [191]. The intermediate particle positions $\boldsymbol{x}^*$ are calculated from the obtained velocity $\boldsymbol{v}^*$ in an explicit manner.

$$\boldsymbol{x}^* = \boldsymbol{x}^n + \boldsymbol{v}^*\Delta t \qquad (2.91)$$

At the intermediate time step, neither the velocity field is divergence-free, nor the fluid incompressible. The divergence-free velocity field is enforced in the corrector step via correction of the intermediate field with the pressure gradient

$$\frac{\boldsymbol{v}^{n+1} - \boldsymbol{v}^*}{\Delta t} = -\frac{1}{\varrho^*}\nabla p^{n+1}. \qquad (2.92)$$

The corrected velocity at time step t^{n+1} is obtained after simple rearrangement to

$$\boldsymbol{v}^{n+1} = \boldsymbol{v}^* - \frac{\Delta t}{\varrho^*}\nabla p^{n+1}. \tag{2.93}$$

However, the pressure term appearing on the right hand side is not known so far. It can be obtained by applying the divergence operator to the correction step shown in equation (2.92). The pressure Poisson equation is obtained to

$$\nabla \cdot \left(\frac{1}{\varrho^*}\nabla p^{n+1}\right) = -\frac{1}{\Delta t}\nabla \cdot \boldsymbol{v}^{n+1} + \frac{1}{\Delta t}\nabla \cdot \boldsymbol{v}^* = \frac{1}{\Delta t}\nabla \cdot \boldsymbol{v}^*, \tag{2.94}$$

with the first summand on the right hand side being zero due to the incompressibility condition. Finally, the corrected particle position $\boldsymbol{x}^{n+1}$ at the full time-step t^{n+1} are calculated with the velocities centered in time at the time steps t^n and t^{n+1}

$$\boldsymbol{x}^{n+1} = \boldsymbol{x}^n + \frac{\boldsymbol{v}^{n+1} + \boldsymbol{v}^n}{2}\Delta t. \tag{2.95}$$

The time integration scheme presented above is of order $O(\Delta t)$ due to the application of the first order explicit Euler scheme in the prediction step. According to Cummins and Rudman, it is possible to enhance the order of the overall scheme by applying higher order integration methods in the prediction step [92]. At the final particle positions obtained from equation (2.95), the flow is incompressible and the velocity field divergence-free within the limits of the truncation error of the spatial discretization scheme. Nonetheless, the error in particle positions and densities is still below the errors appearing in the weakly compressible approach, as it has been shown by Lee et al. as well as Cummins and Rudman [92, 178].

In the projection approach shown above, the incompressibility is enforced via the continuity equation as it is also common in grid-based methods. Yet, a frequently used technique for calculating the fluid density is based on the Sph kernel estimate formalism shown in section 2.1 [56]. This results in the following expression for the fluid density ϱ_i of particle i

$$\varrho_i = \sum_j m_j w(r_{ij}, h). \tag{2.96}$$

In the Moving Particle Semi-Implicit method the latter equation is used as a substitute for the conventional continuity equation [189]. This concept was transfered to Sph by Lo and Shao with minor modifications [166], using the same predictor-corrector scheme but interchanging the right hand side of equation (2.94) as it will be shown subsequently.

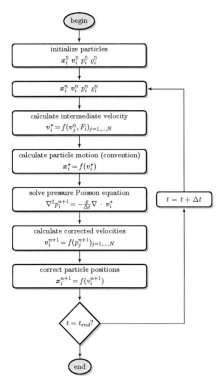

Figure 2.8: Flow sheet of the Incompressible SPH algorithm based a projection method [193].

In general, the fluid is in a compressed state or in tension after the predictor step, having the intermediate density ϱ^*. The idea is to use the deviation of the density at the intermediate step ϱ^* from the incompressible density ϱ^0 to enforce incompressibility. This can be done by discretizing the continuity equation and applying the respective integral boundaries to give

$$\frac{1}{\varrho^0}\frac{\varrho^0 - \varrho^*}{\Delta t} = \nabla \cdot \boldsymbol{v}^*. \tag{2.97}$$

Replacing the divergence of the velocity field in equation (2.94) using the above equation, yields the following expression for the pressure Poisson equation

$$\nabla \cdot \left(\frac{1}{\varrho^*} \nabla p^{n+1} \right) = -\frac{1}{\Delta t^2} \frac{\varrho^* - \varrho^0}{\varrho^0}. \tag{2.98}$$

In the MPS method, the fluid density ϱ_i is interchanged with the particle number density n_i, which is defined as

$$n_i = \sum_j w(r_{ij}, h) \tag{2.99}$$

and used on the right hand side of the pressure Poisson equation, resulting in

$$\nabla \cdot \left(\frac{1}{\varrho^*} \nabla p^{n+1} \right) = -\frac{1}{\Delta t^2} \frac{n^* - n^0}{n^0}. \tag{2.100}$$

Hu and Adams transferred the particle number density based formulation to SPH and found it to be superior for describing multiphase systems with high density differences [169]. In addition, the error is not accumulated using the density based approaches as it is done in the divergence based approach of equation (2.94). However, the density based approaches, shown in equation (2.98) and (2.100), are only partly applicable to free surface flow simulations due to the particle deficit at the free boundary, which can lead to instabilities at free surfaces and pressure fluctuations in general [119, 194–196].

Therefore, combinations of both approaches are proposed by several authors in the literature to enforce incompressibility and the divergence-free condition [24, 170]. Furthermore, Pozorski and Wawrenczuk suggested to simultaneously solve the Poisson equation for density variation and divergence of velocity [197]. Nevertheless, the problems regarding particle clustering, as well as accumulation of the density error are still persisting [170]. The density error accumulation has been effectively suppressed by an approach using the SHAKE scheme, known from constraint molecular dynamics [168]. By iteratively correcting the particle positions and thereby reducing the density error accumulation, a problematic particle clustering tendency is observed using SHAKE, which is reported to be even larger than in the weakly compressible SPH approach by Hu and Adams [170].

3

Physical model and numerical solution

In the present chapter, the governing equations for describing the morphogenesis of open-porous materials are briefly sketched. As described in section 1.2, structure formation is controlled through interactions of compounds with different material behavior and physical states of aggregate. Beginning with the generalized conservation equations, the governing equations of the involved materials are educed. Subsequently, the discretized numerical models for the multiphysics and multiphase system are presented.

3.1 Governing equations

The behavior of a substance on the continuum scale is described by the conservation equations [198, 199]. The following equations are presented in a Lagrangian form due to the Lagrangian nature of the SPH method. In order to estimate an arbitrary physical quantity f of the fluid at time t and position $\boldsymbol{x}$, the value of that quantity needs to be observed along the trajectory from $\boldsymbol{x}(t)$ to $\boldsymbol{x}(t + \delta\,t)$. This can be expressed as

$$\frac{D\,f}{D\,t} := \lim_{\delta\,t \to 0} \frac{f(\boldsymbol{x}(t + \delta\,t), t + \delta\,t) - f(\boldsymbol{x}(t), t)}{\delta\,t}. \tag{3.1}$$

By using a Taylor series expansion, the so called substantial or material time derivative $\frac{D}{Dt}$ can be related to the Eulerian time derivative $\frac{\partial}{\partial t}$ by

$$\frac{D}{Dt} = \frac{\partial}{\partial t} + \frac{\partial}{\partial x^\alpha} \frac{D\, x^\alpha}{Dt} = \frac{\partial}{\partial t} + \boldsymbol{v}^\alpha \cdot \frac{\partial}{\partial x^\alpha}, \tag{3.2}$$

with the velocity vector being represented by $\boldsymbol{v}$ and its components by the Greek
index α. One has to stress, that the Einstein's summation is implied for repeated indices
throughout this work and that the substantial derivative D/Dt coincide with the total
derivative d/dt [61]. Therewith, the continuity equation is formulated as

$$\frac{D\varrho}{Dt} = -\varrho \frac{\partial \boldsymbol{v}^\alpha}{\partial x^\alpha} \tag{3.3}$$

with $\frac{D\varrho}{Dt}$ being the material time derivative of the mass density ϱ.

For an arbitrary material the momentum balances can be written in general form as

$$\varrho \frac{D\boldsymbol{v}^\alpha}{Dt} = \frac{\partial \boldsymbol{\sigma}^{\alpha\beta}}{\partial x^\beta} + \varrho\, \boldsymbol{f} \tag{3.4}$$

with the external body forces per unit mass being abbreviated as $\boldsymbol{f}$. The total stress
tensor $\boldsymbol{\sigma}$ is split up in the isotropic hydrostatic pressure p and the deviatoric stress
tensor $\boldsymbol{\tau}$, characterizing the respective material behavior to give

$$\varrho \frac{D\boldsymbol{v}^\alpha}{Dt} = -\frac{\partial p}{\partial x^\alpha} + \frac{\partial \boldsymbol{\tau}^{\alpha\beta}}{\partial x^\beta} + \varrho\, \boldsymbol{f}. \tag{3.5}$$

Details regarding the models for the stress tensor are given later in the respective
sections.

The thermal energy balance is formulated in terms of the specific internal energy u
for a compressible and reacting material and transformed in the principle variables
temperature T and specific volume v (see appendix I),

$$\begin{aligned}
\varrho\, c_v \frac{DT}{Dt} &+ \varrho \left[T \left. \frac{\partial p}{\partial T} \right|_v - p \right] \frac{Dv}{Dt} = \\
&= -p \frac{\partial \boldsymbol{v}^\alpha}{\partial x^\alpha} - \frac{\partial \dot{\boldsymbol{q}}^\alpha}{\partial x^\alpha} + \boldsymbol{\tau}^{\alpha\beta} \frac{\partial \boldsymbol{v}^\alpha}{\partial x^\beta} - \sum_k MW_k\, u_k \sum_h \nu_{hk}\, r_h.
\end{aligned} \tag{3.6}$$

The specific heat at constant volume is abbreviated as c_v and MW_k depicts the molar
weight of component k, ν_{kh} the stoichiometric coefficient and r_h the reaction rate of

reaction h. For incompressible fluids as well as compressible solids, where thermal expansion is neglected in the present case, the thermal energy balance is simplified to

$$\varrho\, c_v \frac{DT}{Dt} = -\frac{\partial \dot{q}^\alpha}{\partial x^\alpha} + \tau^{\alpha\beta} \frac{\partial v^\alpha}{\partial x^\beta} - \sum_k MW_k\, u_k \sum_h \nu_{hk}\, r_h. \qquad (3.7)$$

For the gas phase, which is treated as an ideal gas, the thermal energy balance is formulated as

$$\varrho\, c_v \frac{DT}{Dt} = -p\frac{\partial v^\alpha}{\partial x^\alpha} - \frac{\partial \dot{q}^\alpha}{\partial x^\beta} + \tau^{\alpha\beta} \frac{\partial v^\alpha}{\partial x^\beta} - \sum_k MW_k\, u_k \sum_h \nu_{hk}\, r_h. \qquad (3.8)$$

The heat flux vector is approximated using Fourier's law

$$\dot{q}^\alpha = -\lambda \frac{\partial T}{\partial x^\alpha}, \qquad (3.9)$$

with the materials heat conductivity displayed by λ.

The material balance for the solute transport of oxygen, denoted by the subscript k, through the polymer matrix reads as follows (see appendix I)

$$\varrho\frac{D\omega_k}{Dt} = -\frac{\partial j_k^\alpha}{\partial x^\alpha} + MW_k \sum_h \nu_{hk} r_h \qquad (3.10)$$

with ω_k being the mass fraction of component k and j_k the diffusive mass flux of component k. The diffusive mass flux j_k follows Fick's first law and is dependent on the mobility parameter b_k^M as shown below

$$j_k^\alpha = -b_k^M\, R\, T\varrho \frac{\partial w_k}{\partial x^\alpha}. \qquad (3.11)$$

In the present work, only the polymer viscosity is dependent on the temperature and is estimated using the Williams-Landel-Ferry model as shown below [200]

$$\eta(T) = \eta_0\, e^{\frac{-C_1(T-T_r)}{C_2+(T-T_r)}}, \qquad (3.12)$$

with η_0 being the reference viscosity and C_1, C_2 and T_r empirical parameters, which are arbitrarily chosen in the present work. For sake of simplicity, the remaining thermodynamic and transport properties are considered constant in the regarded temperature range. However, the future inclusion of temperature dependent material properties is straight forward.

Boundary conditions for multiphase interfacial flow

For the numerical simulation of multiphase systems, the dynamic of the interfaces has to be resolved. While the position of each interface is known initially, its further evolution has to be determined as part of the simulation.

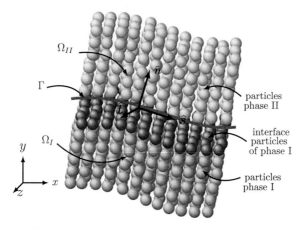

Figure 3.1: Sketch of the internal interface Γ between two phases, labeled phase I and phase II. The surface unit normal vector $\hat{n}$ as well as the tangential vectors $\hat{s}$ and $\hat{t}$ are depicted at the interface.

In the following, a system consisting of two immiscible phases separated by the interface Γ is considered as depicted in figure 3.1. The phases are marked with I and II and possess the material properties ϱ and η amongst others. The conditions the system has to obey at the interface are presented under the assumption of negligible mass transfer between the regarded phases. This results in an identical normal velocity at the interface for both phases, expressed in the so called *kinematic interface condition*

$$v^I \cdot \hat{n}^I + v^{II} \cdot \hat{n}^{II} = 0, \tag{3.13}$$

with $\hat{n}^I$ being the outward pointing interface unit normal vector of the phase I and $\hat{n}^{II}$ the one of phase II, respectively. In addition, the forces at the interface have to be in a state of equilibrium. If one considers material stress and surface tension, the stress balance equation at the interface, the so called *dynamic interface condition*, results in

$$((\boldsymbol{\tau}^I - p^I \boldsymbol{I})\hat{\boldsymbol{n}}^I) \cdot \hat{\boldsymbol{n}}^I + ((\boldsymbol{\tau}^{II} - p^{II} \boldsymbol{I})\hat{\boldsymbol{n}}^{II}) \cdot \hat{\boldsymbol{n}}^{II} + \sigma^{I \leftrightarrow II} \kappa = 0 \qquad (3.14)$$

in normal and

$$((\boldsymbol{\tau}^I - p^I \boldsymbol{I})\hat{\boldsymbol{n}}^I) \cdot \boldsymbol{t}^I - ((\boldsymbol{\tau}^{II} - p^{II} \boldsymbol{I})\hat{\boldsymbol{n}}^{II}) \cdot \boldsymbol{t}^{II} - \nabla_t \sigma^{I \leftrightarrow II} = 0, \qquad (3.15)$$

in tangential direction, with $\hat{\boldsymbol{t}}$ being the unit vector tangential to the interface and orthogonal to $\hat{\boldsymbol{n}}$ in the local Cartesian coordinate system as displayed in figure 3.1. The surface tension term is split up in the normal contribution $\sigma^{I \leftrightarrow II} \kappa$, which depends on the local curvature κ and a tangential part represented by $\nabla_t \sigma^{I \leftrightarrow II}$. In the present work, the influence of surface tension is small in comparison to viscous effects and therefore neglected. This simplifies equation (3.14) to

$$((\boldsymbol{\tau}^I - p^I \boldsymbol{I})\hat{\boldsymbol{n}}^I) \cdot \hat{\boldsymbol{n}}^I = -((\boldsymbol{\tau}^{II} - p^{II} \boldsymbol{I})\hat{\boldsymbol{n}}^{II}) \cdot \hat{\boldsymbol{n}}^{II} \qquad (3.16)$$

and equation (3.15) to

$$((\boldsymbol{\tau}^I - p^I \boldsymbol{I})\hat{\boldsymbol{n}}^I) \cdot \boldsymbol{t}^I = ((\boldsymbol{\tau}^{II} - p^{II} \boldsymbol{I})\hat{\boldsymbol{n}}^{II}) \cdot \boldsymbol{t}^{II}. \qquad (3.17)$$

The normal forces on both sides of the interface possess the same absolute value but with opposite orientation, while the tangential forces exhibit the same absolute value and orientation. Despite the discontinuity in the density over the interface, the total stress tensor (hydrostatic pressure and normal as well as tangential components of the deviatoric stress tensor) is continuous over the interface under these assumptions [201, 202].

In many applications, the evolution of the overall multiphase system is governed only by one particular phase, e.g. when the density ratio between the phases is large. In this case, the actual multiphase system can be simplified by neglecting the influence of the light phase on the heavy phase and the multiphase system is reduced to a single phase system with a free surface as sketched in figure 3.2. In this case, the tangential stress balance (3.17) can be further simplified to

$$(\boldsymbol{\tau}^I \hat{\boldsymbol{n}}^I) \cdot \boldsymbol{t}^I = 0, \qquad (3.18)$$

e.g. if the dynamic viscosity of phase II is small compared to the one of the phase I. Additionally, the normal projection of the dynamic interface condition is also simplified to

$$((\boldsymbol{\tau}^I - p^I \boldsymbol{I})\hat{\boldsymbol{n}}^I) \cdot \hat{\boldsymbol{n}}^I = -p^{II} \cdot \hat{\boldsymbol{n}}^{II}. \qquad (3.19)$$

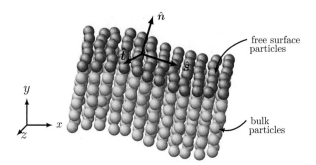

Figure 3.2: Sketch of a free surface or external interface, respectively. The unit normal vector $\hat{n}$ as well as the tangential vectors $\hat{s}$ and $\hat{t}$ to the interface are shown, together with the detected boundary particles.

That simplification will later be used for deriving an hybrid mesh and particle based approach for solving structure formation processes governed by the formation of a gaseous blowing agent and the respective approach is sketched in section 3.2.3. Yet in the following section, the basics of the purely particle based SPH algorithm are depicted first.

3.2 Numerical approach

As mentioned in section 1.2, the structure formation process is governed by the interaction of materials with different physical states of aggregation and properties. In the following section, a numerical approach for multiphysics simulations of multiphase systems based on the Smoothed Particle Hydrodynamics formalism is presented. At first, a newly compressible – incompressible SPH approach, capable of describing the interaction between fluid, solid and gaseous phases, is derived in a general Lagrangian form. Subsequently, the SPH specific details regarding the interpolation and discretization schemes are shown. Since special attention is given to the accurate treatment of interfaces and surfaces, two graded corrected SPH approach are derived. The first approach is based on the Corrected Smoothed Particle method as presented in section 2.5 and improved by using a renormalized kernel function. This approach is especially designed for the simulation of open-porous materials, with a focus on restoring the consistency of the SPH approximation in regions truncated by boundaries. In addition, a simplified variant of the approach is presented as a tradeoff between accuracy and computational effort.

3.2.1 The compressible – incompressible multiphase SPH algorithm

In the following paragraph, the developed SPH approach for treating multiphase materials consisting out of compressible phases and incompressible phases is presented. The approach is composed of a truly incompressible SPH approach and a conventional explicit SPH approach for modeling compressible phases. Thereby, the effective simulation of multiphase systems, consisting out of gaseous, liquid and solid phases, as occurring in structure formation processes governed by the release of a gaseous blowing agent, is achieved. In principle, the description of multiphase systems consisting of compressible and incompressible materials can be approximated by using solely the explicit SPH approach as described in section 2.7.1. However, as already mentioned before, several disadvantages are associated with the explicit SPH approach for treating incompressible materials. The two most prominent ones are the required small time step, which is needed for a stable simulation and the numerical instabilities occurring at evolving and especially at free boundaries. The former is particularly of disadvantage, since rather long-term simulations are needed for the simulation of reaction-induced structure formation processes. The latter is the most important disqualification, since the accurate and stable treatment of free surfaces and interfaces is crucial for the description of evolving open-porous materials. In addition, the inclusion of artificial viscosities and stresses is commonly needed for stabilizing the numerical simulation in the explicit SPH approach.

Hence, a newly compressible – incompressible SPH approach is derived in the present work, which consists of a modified truly incompressible SPH approach for modeling the incompressible phases and a conventional, explicit SPH approach, which is liable for the description of the compressible phases. By coupling both approaches, the modeling of multiphase and multiphysics systems within the SPH framework is achieved without introducing any artificial forces. The sequence of the combined multiphase approach is shown in figure 3.3. In this scheme, the particles are classified in two types, which are either compressible particles labeled by $i \in \Omega_C$ or incompressible particles identified by $i \in \Omega_{IC}$. For the compressible particles $i \in \Omega_C$, the hydrostatic pressure is obtained from an thermodynamic equation of state like the ideal gas law and is therefore an absolute value. In contrast, a relative pressure value is obtained for the incompressible particles, since the pressure value is calculated to assure the incompressibility of the material. In the first step, external forces acting on the particles are summed up regardless of their particle type. Also the deviatoric stress forces are calculated by summation over all neighboring particles, including particles of neighboring phases. The first fork in the pseudocode in figure 3.3 occurs for calculating the hydrostatic pressure and its gradient across the interface. Depending on the type of particle, the pseudocode is either passed through on the right or the left wing, since the absolute pressure values of the incompressible SPH particles are not known and are then estimated by using the dynamic interface condition as introduced in section 3.1. Further details regarding the coupling of two phases are sketched below in paragraph 3.2.1.1. With known pressure, stress tensor as well as external forces, the particle velocities at the intermediate time step t^* are calculated and the particle positions are update accordingly. If a higher order integration scheme, like the Velocity-Verlet integration, is used, the prediction step is repeated for all particles regardless of their specific type as indicated in the sequence in figure 3.3. Afterwards, the integration time step is completed for

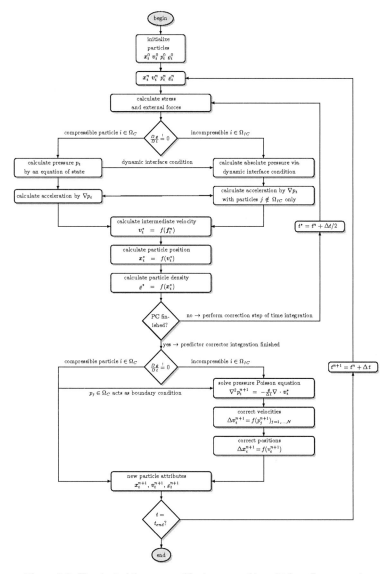

Figure 3.3: Flowchart of the compressible–incompressible multiphase SPH approach.

the compressible particles. In contrast, the incompressibility has to be enforced for incompressible particles. Thereby, the hydrostatic pressure of the compressible particles acts as a boundary condition for solving the pressure Poisson equation. The specific details are shown subsequently in section 3.2.2. With the obtained pressure values, the velocities of the incompressible particles are corrected and their spatial positions are updated. Therewith, the integration time step is completed and the iteration is repeated until the desired final time is reached. In the following paragraph, the details of the spatial and temporal coupling of the multiphase approach are presented.

3.2.1.1 Spatial coupling of the compressible–incompressible SPH algorithm

In the following paragraph, the spatial coupling of interacting phases in the multiphase SPH approach is discussed. For each phase, denoted by phase I and phase II, a different type of particle is used as shown in figure 3.4. For particles in the interior of their respective phase no special treatment is necessary, since they only interact with particles of their own type. However, as soon as the kernel interpolation domain of a particle intersects with particles of a different phase as shown for particle i in figure 3.4, the kinematic interface condition (equation (3.13)) as well as the dynamic interface condition (equations (3.14)–(3.15)) have to be fulfilled at the interface.

Due to its Lagrangian nature, the kinematic condition is naturally satisfied by the SPH method. In addition, the dynamic interface condition, which expresses the continuity of stresses across the interface in the absence of surface tension, is also naturally fulfilled using purely explicit SPH approaches. According to Antoci et al. this is achieved by simply extending the summation in the respective SPH discretization of the stress tensor and momentum conservation equations to all neighboring particles regardless of their type. Hence, the coupling condition of the dynamic interface condition is satisfied by

$$\sigma_i^{\alpha\beta}\hat{n}_i^\beta = -\sigma_j^{\alpha\beta}\hat{n}_j^\beta \quad \text{with } i \in \Omega_I, j \in \Omega_{II}, \tag{3.20}$$

and the no velocity slip boundary condition is satisfied on the contact surface [203]. Yet in the present work, the reaction induced structure formation by release of a blowing agent is modeled. Consequently, the built-up gaseous pressure is the driving force for the material deformation and the absolute value of the gaseous pressure is obtained from the ideal gas equation of state. In contrast, the pressure of the incompressible fluid phase is

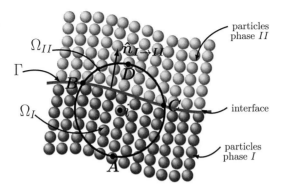

Figure 3.4: Detail of the kernel interpolation domain for the interaction for particle i across the interface. The unit normal vector is depicted as $\hat{n}_{I \to II}$ and the interpolation domain coinciding with phase I is depicted as Ω_I and phase II is labeled as Ω_{II}.

not estimated from a thermodynamic equation of state, but calculated in such a sense to keep the fluid flow divergence free. Therefore, only relative pressure values are obtained for incompressible fluid particles, as well as solid phase particles, where the stress tensor is calculated from the strain rate tensor. Based on these considerations, the conventional approach as used in multiphase WCSPH framework cannot be applied in the respective case and the corresponding absolute pressure value for solid and liquid particles, here marked as phase I particles, in the vicinity of the gaseous phase (phase II), is calculated by using the dynamic interface condition shown in equation (3.16). The absolute pressure of the particle $i \in \Omega_I$ is then calculated under the assumption of negligible surface tension to

$$p_i^{*\,I} = p_j^{II} + ((\tau_i^I - \tau_j^{II})\hat{n}^{I \to II}) \cdot \hat{n}^{I \to II}, \tag{3.21}$$

with p_j^{II} being the absolute pressure of the neighboring particles j of phase II. The estimated absolute pressure $p_i^{*\,I}$ of particle i is then used instead of the relative pressure value in the calculation of the pressure gradient of particle $j \in \Omega_{II}$ interacting with particles of phase Ω_I to give

$$\nabla p_j^{II} = \sum_{k \in \Omega_{II}} V_k(p_j^{II} + p_k^{II})\nabla_k w_{jk} + \sum_{k \in \Omega_I} V_k(p_j^{II} + p_k^{*\,I})\nabla_k w_{jk}. \tag{3.22}$$

So, only absolute pressure values are used in the equation above. In conjunction with the deviatoric stress tensor being calculated by extending the summation of the kernel estimate over the neighboring particles regardless of their type, the continuity of the stress forces over the interface is ensured. Thus, the action of phase I on phase II is covered.

However, the presented approach for calculating absolute pressure values is only valid for interface near particles and cannot be extended to fluid or solid particles lying outside the interface domain. Hence, for the action of phase II on phase I, the interpolation volume has to be divided again into two regions for calculating the pressure gradient of particle $i \in \Omega_I$ across the interface. This is exemplary depicted in figure 3.4, with the volume Ω_I being bounded by the points A, B and C and Ω_{II} by B, C and D. Based on that, the pressure gradient of particle $i \in \Omega_I$ can be split up in

$$\langle \nabla p_i^I \rangle = \int_{\Omega_I} p(\boldsymbol{x}') \nabla w \, d\boldsymbol{x}' + \int_{\Omega_{II}} p(\boldsymbol{x}') \nabla w \, d\boldsymbol{x}'. \tag{3.23}$$

The first integral on the right hand side of equation (3.23) covers the interaction with particles of the same type and can be treated by the SPH gradient approximations as shown in equation (3.22) above or any other gradient formula depicted in section 2.2 and 2.5. But the second integral on the right hand side represents the interactions across the interface. Therein, the pressure gradient is estimated with the absolute pressure of the neighbor phase particle $j \in \Omega_{II}$. For this reason, the discretization used in equation (3.22) is not applicable due to the simultaneous occurrence of relative and absolute pressure values. So, the second term of equation (3.23) has to be discretized using the original SPH gradient operator (equation (2.23)) as follows

$$\int_{\Omega^{II}} p(\boldsymbol{x}') \nabla w \, d\boldsymbol{x}' = \sum_{j \in \Omega_{II}} V_j p_j^{II} \nabla_j w_{ij}, \tag{3.24}$$

with p_j being an absolute pressure of particle $j \in \Omega_{II}$. This results in the overall pressure gradient for the interface particle of phase I formulated as

$$\nabla p_i^I = \sum_{j \in \Omega_I} V_j (p_j^{II} + p_i^I) \nabla_j w_{ij} + \sum_{j \in \Omega_{II}} V_j p_j^{II} \nabla_j w_{ij}. \tag{3.25}$$

One has to stress, that equation (3.24) can also be derived from the *continuum surface force* model introduced by Brackbill for grid based methods and adapted by Morris for particle based approaches, as will be shown later in section 3.2.3 [204, 205]. Again, the dynamic boundary condition is satisfied in conjunction with the deviatoric stress tensor being calculated by extending the summation of the kernel estimate over the neighboring particles regardless of their type. Thus, the influence of phase II exerted on the particles of the opposing phase I is accounted for and the spatial coupling of the two phases is achieved in the prediction step of the projection algorithm. For incompressible particles, the spatial coupling is completed by including the pressure value of the neighbor phase particles as boundary condition for solving the Poisson equation in the projection step. Further details are given in section 3.2.2.1 and 3.2.2.2.

So far, the sequence of the compressible – incompressible SPH approach has been presented, together with the spatial coupling of compressible and incompressible phases. In the subsequent paragraph, the temporal integration of the compressible – incompressible SPH approach is sketched.

3.2.1.2 Time-stepping of the compressible – incompressible SPH algorithm

After presenting the sequence of the compressible – incompressible multiphase SPH approach and the spatial coupling between the regarded phases, the time-stepping algorithm is shown in the following. For illustrative purposes, a first order explicit Euler integration scheme has been used for the Incompressible SPH algorithm shown in section 2.7.2. In order to construct an overall numerical algorithm, which is of second order accuracy in time and space, the Euler integration scheme for the prediction step of the projection method is replaced by a simple two-step predictor-corrector scheme. Predictor-corrector style integrator conserve linear and angular momentum and are commonly used in particle methods [49, 76, 87, 164]. The second order predictor-corrector scheme presented in the following has been introduced by Price and is used for the integration of the prediction step of the projection algorithm (equation (2.90 – 2.91)) as well as for the integration of the equations of motion of the compressible SPH particles [76]. In the following, the superscript indices specify the discrete time step. The prediction step consists of a modified explicit Euler step

$$
\begin{aligned}
v^{n+1/2} &= v^n + \frac{1}{2}\, F^n\, \Delta t \\
x^{n+1/2} &= x^n + \frac{1}{2}\, v^{n+1/2}\, \Delta t \\
T^{n+1/2} &= T^n + \frac{1}{2}\, \dot{T}^n\, \Delta t
\end{aligned}
\tag{3.26}
$$

with the source vector $\boldsymbol{F}^n$ being the right hand side of the respective momentum balance in equation (3.5) evaluated at time step t^n and $\dot{T}^n$ the right hand side of the energy balance (equation (3.6)), respectively.

Once the variables at the intermediate time step are known, the source terms of equation (3.26) are updated using the predicted values at time $t^{n+1/2}$, as it is also done in the symplectic, semi-implicit Euler or Euler-Cromer method [206]. With the updated right hand sides, the corrected values are calculated by

$$
\begin{aligned}
\boldsymbol{v}^{n+1} &= \boldsymbol{v}^n + \boldsymbol{F}^{n+1/2}\,\Delta t \\
\boldsymbol{x}^{n+1} &= \boldsymbol{x}^n + \Delta t\,(\boldsymbol{v}^n + \frac{1}{2}\,\boldsymbol{F}^{n+1/2}\,\Delta t) \\
T^{n+1} &= T^n + \dot{T}^{n+1/2}\,\Delta t.
\end{aligned}
\tag{3.27}
$$

Where applicable, the density or specific volume of the particles, component mass fraction as well as the components of the stress tensor follows the integration scheme used for the temperature. In both, the prediction and correction step, the integration of the spatial position vector uses the updated velocity value $\boldsymbol{v}^{n+1/2}$, resulting in the application of the first and second time derivative for calculating the particle positions. As stated above, the presented integration scheme is used for compressible SPH particles and for the prediction step of the incompressible SPH algorithm. Therefore, the final particle velocity at time step t^{n+1} is not obtained for incompressible particles until the correction with the gradient of the implicitly obtained pressure is accomplished. The final corrected positions of the incompressible particles are centered in time to give

$$
\begin{aligned}
\boldsymbol{v}^{n+1} &= \boldsymbol{v}^{n+1} - \frac{\Delta t}{\varrho}\nabla p^{n+1} \\
\boldsymbol{x}^{n+1} &= \boldsymbol{x}^n + \frac{\boldsymbol{v}^{n+1} + \boldsymbol{v}^n}{2}\Delta t.
\end{aligned}
\tag{3.28}
$$

The velocities and positions of compressible SPH particles obtained after the correction step of the integration scheme (equation (3.27)) do not have to be corrected and are the final values at time step t^{n+1}. As stated above, the overall algorithm is of second order accuracy in space and time due to the second order temporal integration scheme [76]. However, one has to mention, that no significant increase in stability or accuracy has been observed in comparison to the simple explicit Euler integration method for all test cases presented in section 4.

In order to assure a stable temporal integration, the time step Δt is determined via a modified Courant-Friedrichs-Lewy condition. For this reason, it is ensured, that no numerical signal is propagated faster than the maximum velocity occurring in the computational domain in the previous time step t^{n-1} [193, 207]. Following Lo and Shao, the modified CFL-constraint on the time step, which is implemented as shown below, is further increased by using an empirically chosen factor of 0.1 [166]

$$\Delta t \leq 0.1 \frac{L_0}{v_{max}(t^{n-1})}. \tag{3.29}$$

In addition, a second constraint on the time step for the discretization of diffusive parabolic equations is used, which reads [109, 208]

$$\Delta t \leq 0.5 \frac{h^2 \varrho}{\eta}. \tag{3.30}$$

The latter possesses the stronger restriction on the time step, if highly viscous materials are considered [209]. This is especially true, if the simulations are conducted with a high particle resolution, which leads to a small smoothing length due to the smaller inter-particle distances as shown by Ellero and Tanner [106]. Furthermore, an additional third condition is considered for restricting the time step depending on the forces $\boldsymbol{F}_i$ acting on the particles as follows [105]

$$\Delta t \leq min_{\forall i} \frac{h}{|\boldsymbol{F}_i|}. \tag{3.31}$$

The latter condition is especially needed, if the hybrid grid$-$SPH approach is used for modeling the evolution of multiphase systems as it will be shown in section 3.2.3. In overall, to guarantee a stable numerical scheme, the restrictions on the time-step can be quite stringent. This is especially true for the weakly compressible SPH approach, but also holds to a certain extent for the truly incompressible approach. In order to decrease the influence of local effects on the overall allowed time step, Hernquist and Katz introduced an individual time-stepping scheme, which takes the state of each individual particle into account to determine the local time step [84]. The individual particles move hierarchically with their individual time step and the actual overall time is reached after a series of individual sub-steps. However, the approach is not implemented in the present compressible$-$incompressible SPH approach.

3.2.2 Numerical model

As shown in section 2.1, the kernel estimation technique is the foundation of the Smoothed Particle Hydrodynamics method. The kernel estimation of the density of a SPH particle ϱ_i reduces to the simple expression

$$\varrho_i = \sum_j m_j w_{ij}. \tag{3.32}$$

Instead of solving the continuity equation, the density of particle i is calculated by a simple summation over the neighboring particle masses m_j. If one takes the substantial time derivative of equation (3.32), one obtains the following expression under the presumption of a constant particle mass and smoothing length h

$$
\begin{aligned}
\frac{D\varrho_i}{Dt} &= \sum_j m_j \frac{Dw_{ij}}{Dt} \\
&= \sum_j m_j \left(\frac{\partial w_{ij}}{\partial t} + \frac{Dr_{ij}}{Dt} \frac{\partial w_{ij}}{\partial r_{ij}} \right) \\
&= \sum_j m_j \left(\frac{\partial r_{ij}}{\partial \boldsymbol{x}_i} \frac{D\boldsymbol{x}_i}{Dt} + \frac{\partial r_{ij}}{\partial \boldsymbol{x}_j} \frac{D\boldsymbol{x}_j}{Dt} \right) \frac{\partial w_{ij}}{\partial r_{ij}} \\
&= \sum_j m_j \left(\nabla_i r_{ij} \cdot \boldsymbol{v}_i + \nabla_j r_{ij} \cdot \boldsymbol{v}_j \right) \frac{\partial w_{ij}}{\partial r_{ij}} \\
&= \sum_j m_j \left(\boldsymbol{v}_j - \boldsymbol{v}_i \right) \frac{\boldsymbol{x}_j - \boldsymbol{x}_i}{r_{ij}} \frac{\partial w_{ij}}{\partial r_{ij}} \\
&= \sum_j m_j \left(\boldsymbol{v}_j - \boldsymbol{v}_i \right) \nabla_j w_{ij}.
\end{aligned}
\tag{3.33}
$$

Using the SPH gradient operator shown in equation (2.39), the above equation can be transferred back to the continuous description of the continuity equation

$$\frac{D\varrho}{Dt} = -\varrho \, \nabla \cdot \boldsymbol{v}. \tag{3.34}$$

The presented approach, which was taken from the work of Price [76], indicates that both expressions, equation (3.32) and (3.33), are SPH discretized versions for the continuity equation. And both representations have their advantages. The advantage of evaluating the density by interpolation from neighboring masses, as shown in equation (3.32), lies in its implicit mass conservation [49, 50]. However, the expression is not suited for modeling multiphase systems with sharp density discontinuities or free surfaces.

For the simulation of multiphase systems, the integration of the continuity equation as shown in equation (3.33) is mostly used, but at the expense of losing the implicit mass conservation. Besides, the density error introduced in every time step is summed up over time [169, 193]. In addition, the latter approach is more prone to particle clustering, driving the need for reinitializing or repositioning particles during the simulation [81, 91, 210]. Therefore, the choice of the discretization scheme is strongly application dependent, with equation (3.32) being mainly used in single phase flow.

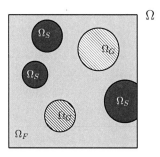

Figure 3.5: Computational domain Ω composed of different materials, denoted by Ω_S for solid, Ω_F for liquid or incompressible fluid and Ω_G for gaseous phases.

In the presented work, where the accurate treatment of interfaces and free surfaces is crucial, a combination of both approaches is applied. The density of particle i is evaluated by interpolation with the neighboring particles as shown in equation (3.32). However, in order to resolve the density discontinuity, only particles j belonging to the same phase $j \in \Omega_i$ are included in the summation. Since there are no mass contributions from neighboring particles of the opposing phase, large density differences between the respective particles can be described. The CSPM approach, as presented in section 2.5, is used for the kernel estimate, resulting in a renormalized kernel or Shepard kernel as displayed in equation (2.63). This results in the following expression for the density

$$\varrho_i = \sum_{j \in \Omega_i} m_j \widehat{w}_{ij}, \qquad (3.35)$$

with the Shepard kernel being defined as

$$\widehat{w}_{ij} = \frac{w_{ij}}{\sum_{k \in \Omega_i} V_k w_{ik}}. \qquad (3.36)$$

In most SPH approaches, the volume of a particle V_i is estimated from the given mass and its density to give

$$V_i = \frac{m_i}{\varrho_i}. \tag{3.37}$$

Yet, in the presented approach this direct link is not used and the temporal evolution of the particle volume is calculated from the continuity equation. On these grounds, the particle density ϱ_i is replaced with the specific volume v_i in the equation of continuity

$$\frac{D \ln \frac{\varrho_i^0}{\varrho_i}}{Dt} = \frac{D \ln \frac{v_i}{v_i^0}}{Dt} = \nabla \cdot \boldsymbol{v}, \tag{3.38}$$

with v_i^0 and ϱ_i^0 being the initial specific particle volume, respectively density. Due to possibly low particle volumes and to assure its positiveness, the continuity equation is written in a logarithmic form, since fast changes in the volume are not expected in the present application. To calculate the particles specific volume v_i, no distinction between the different phases is made, which results in the summation over all the neighboring particles in the discretized form of the continuity equation. Thereby, the connection between the phases is restored, while keeping the sharp discontinuity in the density field across the interface. The presented idea was first published by Grenier *et al.* in the context of multiphase flow using the weakly compressible SPH approach and is extended to Incompressible SPH in the work at hand [119, 211].

In the following, a model for simulating multiphase flow with compressible and incompressible phases as well as free boundaries is introduced. The multiphase system is composed of solid, incompressible liquid and gaseous phases as shown in figure 3.5. As stated in section 2.3, the occurrence of interfaces and free boundaries, which truncate the interpolation domain of the kernel estimate, dramatically deteriorate the accuracy of the interpolation. Based on that, corrected SPH discretization schemes are needed for the quantitative description of structure forming process. Due to the higher computational demand associated with the correction algorithms, two models with graduated degree of correction are proposed. In the first model, a renormalized kernel function is used together with the CSPM approach presented in section 2.5. In the second model a renormalized kernel is used also, but the appearing derivatives are only corrected by the constant denominator of the Shepard kernel. The second model disclaims the evaluation of the correction matrix for the derivatives (equation (2.67)) and is for this reason computationally less expensive at the cost of a reduced accuracy. One has to stress, that both models are based on the same pseudocode scheme as described above, but different SPH approximations and discretization techniques are used. In the subsequent two sections, both models are presented in detail.

3.2.2.1 Normalized Corrected SPH model

In the following paragraph, the discretization of the conditional equations of section 3.1 using the Normalized Corrected SPH (NCSPH) method is presented for a multiphase system consisting of solid, incompressible liquid and gaseous phases as shown in figure 3.5. The NCSPH approach is based on the Corrected Smoothed Particle method of Chen *et al.* and further developed by using a renormalized kernel function. Therewith, the consistency of the SPH approximation in regions truncated by a boundary is restored, making the approach perfectly suited for the simulation of the morphogenesis of open-porous materials.

Discretization of the continuity equation

As already shown above, a renormalized kernel or Shepard kernel is used for the kernel estimate. By using a constant kernel correction, which results in

$$\widehat{w}_{ij} = \frac{w_{ij}}{\sum_k V_k w_{ik}}, \tag{3.39}$$

the normalization condition, shown in equation (2.4) is satisfied on the whole domain Thus, constant functions are exactly evaluated and the interpolation is strongly improved at domain boundaries and for irregular particle distributions. One has to stress, that it is in principle possible to construct kernel functions, which exactly interpolate polynomials up to any degree as used in the Moving Least-Square Kernel Galerkin method [157, 158]. However, for linear corrected kernel functions for example, one additional space depended parameter has to be computed and the evaluation of the gradient of the kernel function is troublesome and computationally extraordinary costly as stated by Bonet and Lok [87]. That is why, only the constant kernel correction is used in the present work. As already shown above (equation (3.35)), the discretized continuity equation is based on the Shepard kernel interpolation and is repeated for sake of completeness,

$$\varrho_i = \sum_{j \in \Omega_i} m_j \widehat{w}_{ij}^{\Omega_i} = \frac{\sum_{j \in \Omega_i} m_j w_{ij}}{\sum_{k \in \Omega_i} V_k w_{ik}}. \tag{3.40}$$

On has to stress, that the summation over the neighboring particles j in the numerator as well as in the Shepard kernel is limited to particles of their own phase as indicated by $\widehat{w}_{ij}^{\Omega_i}$. Thereby, the discontinuity in the density field over the interface is explicitly resolved.

In order to discretize the conditional equation for the particle volume, shown in equation (3.38), the first order derivative of the normalized kernel is calculated to

$$\nabla_i \widehat{w}_{ij} = \frac{\nabla_i w_{ij} - w_{ij} \frac{\sum_k V_k \nabla_i w_{ik}}{\sum_k V_k w_{ik}}}{\sum_k V_k w_{ik}}. \tag{3.41}$$

By using a renormalized kernel, the condition for the first moment of the kernel function $\sum_j \nabla_i \widehat{w}_{ij} = 0$ is satisfied on the whole interpolation domain. Thereby, constant functions are exactly reproduced. For exact reproduction of linear functions, the gradient has to be corrected as shown in equation (2.67). With the renormalized kernel function $\widehat{w}_{ij}$, the correction matrix $\boldsymbol{A}$ is modified to

$$\boldsymbol{A} = -\sum_j V_j \nabla_i \widehat{w}_{ij} \otimes (\boldsymbol{x}_j - \boldsymbol{x}_i), \tag{3.42}$$

following in the corrected gradient operator as given by Bonet and Lok [87]

$$\widetilde{\nabla}_i \widehat{w}_{ij} = \left(-\sum_j V_j \nabla_i \widehat{w}_{ij} \otimes (\boldsymbol{x}_j - \boldsymbol{x}_i) \right)^{-1} \nabla_i \widehat{w}_{ij}, \tag{3.43}$$

where $\widetilde{\nabla}_i \widehat{w}_{ij}$ is the corrected first order derivative of the renormalized kernel $\widehat{w}_{ij}$. With the corrected gradient of the normalized kernel, the corrected first order derivative of a field function f can be expressed as

$$\langle \nabla f \rangle_i = \sum_j V_j (f_j - f_i) \cdot \widetilde{\nabla}_i \widehat{w}_{ij}. \tag{3.44}$$

One has to note, that the presented formalism is based on the CSPM shown in section 2.5. However, in the conventional CSPM approach a conventional kernel without renormalization is used, which is a clear disadvantage as will be seen later. With the presented corrected gradient formula, the conditional equation for the particle volume V_i is discretized to

$$\frac{D \ln(V_i/V_i^0)}{Dt} = \sum_j V_j (\boldsymbol{v}_j - \boldsymbol{v}_i) \cdot \widetilde{\nabla}_i \widehat{w}_{ij}, \tag{3.45}$$

where the summation is enlarged over the full particle neighborhood regardless of the particle type.

Discretization of the momentum balance

In order to naturally satisfy the conservation principles, the discretization of the stress forces in the momentum and energy equation has to be consistent with the discretization of the continuity equation [87]. A consistent discretization scheme for the corrected and normalized gradient of the stress tensor is derived in the following by using the variational approach presented in section 2.6. However, due to kernel renormalization and correction, the kernel function lost its symmetric property. Hence, the former constraint used in the derivations in section 2.6

$$\widetilde{\nabla}_i \widehat{w}_{ij} \neq -\widetilde{\nabla}_j \widehat{w}_{ij} \tag{3.46}$$

is not valid any more. Therefore, a new discretization scheme for the total stress forces $\boldsymbol{\sigma}$ evolves from equation (2.75) and reads

$$\langle \nabla \cdot \boldsymbol{\sigma} \rangle_i = \sum_j V_j \big(\boldsymbol{\sigma}_i \cdot \widetilde{\nabla}_i \widehat{w}_{ij} - \boldsymbol{\sigma}_j \cdot \widetilde{\nabla}_j \widehat{w}_{ij} \big). \tag{3.47}$$

The new discretization scheme is clearly different from the conventional one, depicted in equation (2.47), and inherently conserves linear and angular momentum due to its derivation from a variational principle. But the presented scheme proved to be instable and insufficiently accurate for discretization of the stress forces, since it is not zeroth order consistent (i.e. constant functions are not reproduced exactly). Nevertheless, the explicit conservation of momentum is a remarkable feature for a corrected kernel gradient, which will be exploited for the construction of the second order derivative used for the construction of the Laplace matrix in the pressure correction step as it will be shown later.

Due to its inconsistency, equation (3.47) above will not be used to discretize the stress forces in the momentum and energy balance. Thus, the conventional gradient shown in equation (2.47) will be enhanced by replacing the conventional gradient of the kernel with the corrected and renormalized version $\widetilde{\nabla}_i \widehat{w}_{ij}$ to give

$$\langle \nabla \cdot \boldsymbol{\sigma} \rangle_i = \sum_j V_j \left(\boldsymbol{\sigma}_i + \boldsymbol{\sigma}_j \right) \cdot \widetilde{\nabla}_i \widehat{w}_{ij}. \tag{3.48}$$

By using $\widetilde{\nabla}_i \widehat{w}_{ij}$ the natural conservation of momentum due to the symmetry between particle pairs is lost as stressed above. Anyway, the conservation principles are satisfied regarding the whole domain as will be shown later. In addition, equation (3.48) reproduces constant and linear polynomials exactly on the whole domain due to the renormalized kernel and its corrected gradient.

With the presented discretization, the momentum balance of each particle can be formulated. However, as shown in section 3.2.1, one has to distinguish between solid phase as well as incompressible fluid phase particles $i \in \Omega_S \cup \Omega_F$ and compressible gas phase particles $i \in \Omega_G$, where the absolute pressure of the latter is estimated using a thermodynamic equation of state. Regarding these compressible particles $i \in \Omega_G$, the momentum balance equation is formulated as

$$
\begin{aligned}
\varrho_i \frac{D\,v_i}{D\,t} = &\sum_j V_j(\tau_i + \tau_j) \cdot \widetilde{\nabla}_i \widehat{w}_{ij} - \sum_{j \in \Omega_G} V_j(p_i + p_j)\widetilde{\nabla}_i \widehat{w}_{ij} \\
&- \sum_{j \in \Omega_F, \cup \Omega_S} V_j(p_i + p_j^*)\widetilde{\nabla}_i \widehat{w}_{ij} + \varrho_i f_i \quad \forall i \in \Omega_G.
\end{aligned}
\tag{3.49}
$$

For the compressible particles $i \in \Omega_G$, the system of differential equations is closed with the equation of state, e.g. of an ideal gas

$$
p_i = \frac{\varrho_i R T_i}{MW},
\tag{3.50}
$$

where R is the universal gas constant and MW the molecular weight of the gas mixture. In contrast, the corresponding (absolute) pressure values for the interacting incompressible fluid and compressible solid particles have to be estimated using the dynamic interface condition as stated in section 3.1. Therewith, the estimated absolute pressure is calculated for these particles following equation (3.21)

$$
p_j^* = p_i + ((\tau_j - \tau_i)\hat{n}) \cdot \hat{n} \qquad \forall\, i \in \Omega_G, \; j \in \Omega_F \cup \Omega_S.
\tag{3.51}
$$

One has to stress, that the distinction in the pressure term of equation (3.49) between fluid and solid particles as well as compressible gas phase particles is only motivated by preventing a mixture of absolute and relative pressure values for the different types of particles. Furthermore, the deviatoric stress tensor τ of the solid particles included in the summation of the stress gradient terms of equation (3.49) is estimated from the total stress tensor of the solid particles j obeying Hook's law

$$
\sigma_j^{\alpha\beta} = 2\mu\epsilon_j^{\alpha\beta} + \lambda\delta^{\alpha\beta}\epsilon_j^{\gamma\gamma}
\tag{3.52}
$$

as follows

$$
\tau_j^{\alpha\beta} = \sigma_j^{\alpha\beta} + \delta^{\alpha\beta}p_j = \sigma_j^{\alpha\beta} - \delta^{\alpha\beta}\left(\frac{2}{3}\mu + \lambda\right)\epsilon_j^{\gamma\gamma} \qquad \forall j \in \Omega_S.
\tag{3.53}
$$

Further details of Hook's constitutive equation are given in section 4.2.4.

For gaseous and liquid particles, the deviatoric stress part of the dynamic interface condition is naturally met by expanding the summation over the complete particle neighborhood regardless of the particle type for calculating the deviatoric stress term $\boldsymbol{\tau}$. If gaseous or liquid particles interact with solid particles in the calculation of the deviatoric stress term, the respective material behavior of the fluid particles is imposed on the solid particles for that particular interaction.

As stated above, the incompressible fluid is modeled using a projection based approach similar to the one shown in section 2.7.2. For this reason, the momentum balance for incompressible particles looks slightly different. And since it has been found by Cummins and Rudman, that a full projection[1] yields more accurate and stable results [92], the momentum balance for incompressible fluid particles $i \in \Omega_F$ is reduced to

$$
\begin{aligned}
\varrho_i \frac{D \, \boldsymbol{v}_i}{D \, t} = &\sum_{j \in \Omega_G \cup \Omega_F} V_j (\boldsymbol{\tau}_i + \boldsymbol{\tau}_j) \cdot \widetilde{\nabla}_i \widehat{w}_{ij} + \sum_{j \in \Omega_S} V_j (\boldsymbol{\sigma}_i + \boldsymbol{\sigma}_j) \cdot \widetilde{\nabla}_i \widehat{w}_{ij} \\
&- \sum_{j \in \Omega_G} V_j p_j \widetilde{\nabla}_i \widehat{w}_{ij} + \varrho_i \boldsymbol{f}_i \qquad \forall i \in \Omega_F,
\end{aligned}
\tag{3.54}
$$

where the special treatment for the pressure gradient term across the interface between gaseous and liquid particles is employed as described in section 3.2.1.1 to satisfy the dynamic interface condition between the regarded phases. Hence, the summation index of the pressure gradient term is reduced to particles belonging to compressible gas phases Ω_G only and the pressure gradients between the incompressible particles is neglected, since these are evaluated in the correction step of the projection method. For fluid particles interacting with solid ones, the total stress tensor $\boldsymbol{\sigma}$ is applied in the gradient term.

For solid particles $i \in \Omega_S$, the momentum balance looks similar to the balance equation for incompressible fluid particles

$$
\begin{aligned}
\varrho_i \frac{D \, \boldsymbol{v}_i}{D \, t} = &\sum_{j \in \Omega_S \cup \Omega_F} V_j (\boldsymbol{\sigma}_i + \boldsymbol{\sigma}_j) \cdot \widetilde{\nabla}_i \widehat{w}_{ij} + \sum_{j \in \Omega_G} V_j (\boldsymbol{\tau}_i + \boldsymbol{\tau}_j) \cdot \widetilde{\nabla}_i \widehat{w}_{ij} \\
&- \sum_{j \in \Omega_G} V_j p_j \widetilde{\nabla}_i \widehat{w}_{ij} + \varrho_i \boldsymbol{f}_i \qquad \forall i \in \Omega_S,
\end{aligned}
\tag{3.55}
$$

---

[1]In a full projection, the term of the pressure gradient of the previous time step is neglected in the prediction step.

with the exception, that the total stress tensor $\boldsymbol{\sigma}$ is applied in the interaction with fluid particles also. For evaluating the stress of solid particles, which possess fluid particles in their neighborhood, the material law of the fluid particles is imposed on the solid particles for that particular interaction as stated also above.

However, if considering multiphase systems with large density differences between the respective phase, the first order derivatives in the equations (3.49), (3.54) and (3.55) are sometimes modified to increase stability. According to Hu and Adams, an inter-particle averaged expression for $\frac{\nabla p}{\varrho}$, $\frac{\nabla \tau^{\alpha,\beta}}{\varrho}$ and $\frac{\nabla \sigma^{\alpha,\beta}}{\varrho}$ should be used to ensure numerical stability and therewith accuracy [169, 170]. This is due to the fact, that the continuity of the total stress tensor $\boldsymbol{\sigma}$ and the derivative $\frac{\nabla \sigma^{\alpha,\beta}}{\varrho}$ has to be ensured across the interface, even though the density field is discontinuous. The inter-particle averaged expression for the pressure gradient is exemplary below

$$\langle \nabla p \rangle_i = \sum_j V_j \frac{2(\varrho_j p_i + \varrho_i p_j)}{\varrho_i + \varrho_j} \widetilde{\nabla}_i \widehat{w}_{ij}. \tag{3.56}$$

Recently, it has been reported by Monaghan, that an inter-particle averaging is not necessary for multiphase simulations with high density ratios between the phases as long as the particle spacing is small enough [212]. Though, Monaghan applied an artificial surface tension for the interaction of particles of different phases. The artificial surface tension term has also been used previously by Grenier *et al.* as well as Monaghan himself [119, 213].

As already stated above, the dynamic interface condition is satisfied for gaseous and liquid particles by expanding the summation for calculating the respective deviatoric stress term $\boldsymbol{\tau}$ over the complete particle neighborhood. In case of solid particles, the gradient of the total stress tensor $\boldsymbol{\sigma}$ is already used in the balance equations. Details regarding the material models and the discretization of the total, respectively deviatoric stress tensor are given in the particular sections, since several different material models are used in the present work.

After all particles are moved according to the time integration scheme consisting out of equation (3.26) and (3.27), the incompressible flow condition has to be enforced for the incompressible particles using the projection approach. Therefore, the following pressure Poisson equation has to be solved.

$$\nabla \cdot \left(\frac{1}{\varrho^*} \nabla p^{n+1} \right) = \frac{1}{\Delta t} \nabla \cdot \boldsymbol{v}^*, \tag{3.57}$$

The divergence of the velocity field, as shown on the right hand side of equation (3.57) is calculated as

$$\nabla \cdot \boldsymbol{v}_i^* = \sum_{j \in \Omega_F} V_j (\boldsymbol{v}_j^* - \boldsymbol{v}_i^*) \widetilde{\nabla}_i \widehat{w}_{ij}^{\Omega_F}, \tag{3.58}$$

with only incompressible particles included in the summation. Therewith, constant and linear polynomials are reproduced exactly on the respective domain.

In general, the Laplacian in the pressure Poisson equation (3.57) can be formulated by using the second order derivative formulated with the second order derivative of the kernel function, as shown in equation (2.51). However, that approach is known to be very sensitive to particle disorder [20]. In addition, pressure fluctuations can arise, since the second order derivative of the kernel function changes its sign when undercutting a certain particle distance as previously depicted in figure 2.3. This observation was also made when using the corrected second order derivative of the CSPM with a renormalized kernel function as depicted in appendix E. Furthermore, it was found by Cummins and Rudman, that an exact projection leads to the distinct pressure decoupling pattern, which has also been observed on unstaggered finite difference grids due to the co-location of pressure and velocity [92, 214]. During the course of the simulation, these effects can lead to pressure oscillations and even stop the simulation. For this reason, an approximate projection operator is used in the conventional ISPH approach to avoid this problem [92]. But if using a renormalized kernel and the kernel gradient correction, the commonly used Laplacian, shown in equation (2.55), leads to a non-symmetric coefficient matrix of the linear equation system. The non-symmetric matrix is in general more difficult and costly to solve than the symmetric positive definite matrix occurring in the conventional ISPH approach. Hence, a new Laplacian operator is derived for the corrected approach. The approach is based on the corrected first order derivative shown in equation (3.47), which is deduced from a variational principle and a finite difference approximation of the first order derivative. The corrected SPH formulation of the first order derivative of the pressure gradient term simplifies to

$$\nabla \cdot \left(\frac{1}{\varrho} \nabla p \right)_i = \sum_j V_j \left(\frac{\nabla_i p_i}{\varrho_i} \widetilde{\nabla}_i \widehat{w}_{ij} - \frac{\nabla_j p_j}{\varrho_j} \widetilde{\nabla}_j \widehat{w}_{ij} \right). \tag{3.59}$$

The finite difference approximation of the gradient between two particles transferred in the Cartesian plane is given by

$$\left(\frac{\partial p}{\partial \boldsymbol{r}}\right) = \left(\frac{\partial p}{\partial \boldsymbol{x}}\right)\left(\frac{\partial \boldsymbol{x}}{\partial \boldsymbol{r}}\right), \tag{3.60}$$

which results in

$$\left(\frac{\partial p_i}{\partial \boldsymbol{x}_i}\right) = \frac{p_j - p_i}{r_{ij}}\frac{\boldsymbol{x}_j - \boldsymbol{x}_i}{r_{ij}} \tag{3.61}$$

Inserting the latter equation in equation (3.59) results in the new corrected SPH discretization for the Laplacian operator

$$\nabla \cdot \left(\frac{1}{\varrho}\nabla p\right)_i = \sum_{j\in\Omega_F} V_j \frac{(p_j - p_i)}{r_{ij}^2}(\boldsymbol{x}_j - \boldsymbol{x}_i) \cdot \left(\frac{1}{\varrho_i}\widetilde{\nabla}_i \widehat{w}_{ij} - \frac{1}{\varrho_j}\widetilde{\nabla}_j \widehat{w}_{ij}\right), \tag{3.62}$$

with only incompressible neighboring particles $j \in \Omega_F$ included in the summation. This implicates a system of linear equations following

$$\sum_{j\in\Omega_F} a_{ij}p_j = b_i \quad \forall\, i \in \Omega_F, \tag{3.63}$$

where the components of the matrix are computed by

$$a_{ij} = V_j \frac{(\boldsymbol{x}_j - \boldsymbol{x}_i) \cdot (\frac{1}{\varrho_i}\widetilde{\nabla}_i \widehat{w}_{ij} - \frac{1}{\varrho_j}\widetilde{\nabla}_j \widehat{w}_{ij})}{r_{ij}^2} \quad \forall\, j \neq i \cup j \in \Omega_F, \tag{3.64}$$

$$a_{ii} = -\sum_{j\neq i} a_{ij}. \tag{3.65}$$

The right hand side of equation (3.63) is discretized to

$$b_i = \sum_{j\in\Omega_F} V_j(\boldsymbol{v}_j^* - \boldsymbol{v}_i^*) \cdot \widetilde{\nabla}_i \widehat{w}_{ij}^{\Omega_F} \tag{3.66}$$

for each particle $i \in \Omega_F$. The asterik $*$ at the velocity vector $\boldsymbol{v}$ marks the velocity of the particle after the predictor step as described in section 2.7.2 and 3.2.1. To guarantee a symmetric and positive definite coefficient matrix over the whole simulation time, the particle volume V_j in equation (3.64) should be replaced by the respective harmonic average

$$V_{ij} = \frac{2V_iV_j}{(V_i + V_j)}. \tag{3.67}$$

By using this approach, the resulting symmetric and positive definite coefficient matrix of the linear system can be solved with the fast conjugate gradient solver [127, 215]. In conjunction with an Incomplete Cholesky preconditioner, the calculation is stable and robust even for strongly distorted particle configurations as it has been also reported by Koshizuka and coworker [193].

For multiphase systems, where the incompressible phase interacts with other compressible phases, equation (3.64) to (3.66) have to be amended. At the interface, the von Neumann boundary condition for the pressure is applied. The boundary condition is derived from the Poisson equation shown in equation (3.57) by using Gauß theorem, which results in

$$\nabla p \cdot \hat{\boldsymbol{n}} = \frac{\varrho}{\Delta t} \boldsymbol{v}^* \cdot \hat{\boldsymbol{n}}. \tag{3.68}$$

The pressure gradient is discretized by

$$\nabla p_i = \sum_j V_j \left(p_i + p_j \right) \cdot \widetilde{\nabla}_i \widehat{w}_{ij}, \tag{3.69}$$

with the summation being extended over all particles regardless of their type. Based on that, the entry on the principal diagonal of the coefficient matrix has to be complemented for incompressible particles $i \in \Omega_F$ having compressible particles of phase Ω_G and Ω_S in their interpolation domain. So, equation (3.64) is further developed to

$$a_{ij} = V_j \frac{(\boldsymbol{x}_j - \boldsymbol{x}_i) \cdot \left(\frac{1}{\varrho_i} \widetilde{\nabla}_i \widehat{w}_{ij} - \frac{1}{\varrho_j} \widetilde{\nabla}_j \widehat{w}_{ij} \right)}{r_{ij}^2} + V_j \widetilde{\nabla}_i \widehat{w}_{ij} \cdot \hat{\boldsymbol{n}} \quad \forall \, j \neq i, \tag{3.70}$$

$$a_{ii} = -\sum_{\substack{j \in \Omega_F \\ j \neq i}} a_{ij} + \sum_{k \notin \Omega_F} V_k \widetilde{\nabla}_i \widehat{w}_{ik} \cdot \hat{\boldsymbol{n}}. \tag{3.71}$$

Moreover, the right hand side of the Poisson equation of the incompressible particle $i \in \Omega_F$ has to be modified to

$$\begin{aligned} b_i = & \sum_{j \in \Omega_F} V_j (\boldsymbol{v}_j^* - \boldsymbol{v}_i^*) \cdot \widetilde{\nabla}_i \widehat{w}_{ij}^{\Omega_F} \\ & - \frac{\varrho_i}{\Delta t} (\boldsymbol{v}_i \cdot \hat{\boldsymbol{n}}_i) - \sum_{j \notin \Omega_F} V_j p_j \widetilde{\nabla}_i \widehat{w}_{ij} \cdot \hat{\boldsymbol{n}}_i, \end{aligned} \tag{3.72}$$

with the last term being the contribution of the compressible particles $j \notin \Omega_F$ to the pressure gradient of particle i.

For the computation of the surface normal vector of particle i $\boldsymbol{n}_i$, a so called color function χ is introduced as it has been done by Morris, which is defined as

$$\chi_i = \begin{cases} 0 & \forall i \in \Omega_G \\ 1 & \forall i \in \Omega_F \\ 2 & \forall i \in \Omega_S \end{cases} \tag{3.73}$$

Based on that, the normal vector to the interface is calculated to

$$n_i = \frac{1}{\Delta \chi_{ij}} \sum_j V_j (\chi_j - \chi_i) \nabla_i w_{ij}, \qquad (3.74)$$

with χ_i being the color function of particle i and $\Delta \chi_{ij}$ is defined as the leap in the color functions across the interface defined as $\Delta \chi_{ij} = |\chi_j - \chi_i|$ [205]. Since only the normal vector to the liquid surface is of concern, the calculation can be simplified to

$$n_i = \sum_{j \in \Omega_F} V_j \nabla_i w_{ij}. \qquad (3.75)$$

The unit normal vector of particle i to the surface $\hat{n}_i$ is then computed to

$$\hat{n}_i = \frac{n_i}{|n_i|} \qquad (3.76)$$

With the obtained pressure field, the velocities and positions of the incompressible particles have to be corrected. The respective pressure gradient is expressed by the non-symmetric gradient version, depicted in equation (3.48), as

$$\nabla p_i = \sum_{j \in \Omega_F} V_j (p_i + p_j) \widetilde{\nabla}_i \widehat{w}_{ij}, \qquad (3.77)$$

where the summation is limited to incompressible fluid particles only. One has to note, that the latter equation results from equation (3.48) under the assumption of an even kernel function. In contrast to the conventional, uncorrected gradient formula, the self-contribution of the respective particle has to be considered, since the gradient of the corrected kernel function does not vanish with the inter-particle distance r_{ij} going to zero. One has to stress, that this has to be considered for all corrected SPH gradient operators. This results in the following expression

$$\nabla_i \widehat{w}(r_{ii}) = \frac{-w(r_{ii}) \frac{\sum_k V_k \nabla_i w_{ik}}{\sum_k V_k w_{ik}}}{\sum_k V_k w_{ik}}, \qquad (3.78)$$

which in turns has to be multiplied by the inverse of equation (2.68) to obtain the fully corrected contribution $\widetilde{\nabla}_i \widehat{w}(r_{ii})$ of particle i. Therewith, the corrected gradient of the pressure is calculated to

$$\nabla p_i = 2 \, p_i \widetilde{\nabla}_i \widehat{w}(r_{ii}) + \sum_{j \neq i} V_j (p_i + p_j) \widetilde{\nabla}_i \widehat{w}_{ij}. \qquad (3.79)$$

88

A remarkable feature of the above formulation lies in the exact reproduction of constant and linear functions due to the corrected gradient and renormalized kernel $\widetilde{\nabla}\widehat{w}_{ij}$. Without the applied correction as it is used in the Chen's Corrected Smoothed Particle method (section 2.5), the original form of the gradient operator is not even zeroth order complete. This is also true for the conventional gradient operator, shown in equation (2.47), as it has been demonstrated by Colin *et al.* [190].

Discretization of the energy balance

In the following paragraph, the discretization of the thermal energy balance shown in equation (3.6) is discussed, with Fourier's law being used for modeling heat conduction. With the corrected normalized SPH formalism, the energy balance is discretized to

$$
\begin{aligned}
\varrho_i c_{v,i} \frac{D\,T_i}{D\,t} = -p_i \sum_j V_j (\boldsymbol{v}_j - \boldsymbol{v}_i) \cdot \widetilde{\nabla}_i \widehat{w}_{ij} + \left\langle \frac{\partial}{\partial \boldsymbol{x}^\alpha} \left(\lambda_i \frac{\partial T}{\partial \boldsymbol{x}^\alpha} \right) \right\rangle_i \\
- \langle \boldsymbol{\tau} \rangle_i : \sum_j V_j (\boldsymbol{v}_j - \boldsymbol{v}_i) \widetilde{\nabla}_i \widehat{w}_{ij} - \sum_k M W_k\, u_{k,i} \sum_h \nu_{hk}\, r_{h,i},
\end{aligned}
\tag{3.80}
$$

with : denoting the double-dot scalar product of two tensors. For incompressible liquid particles as well as compressible solid particles, the first term on the right hand side of the thermal energy balance is neglected. The heat conduction term is discretized using a corrected variant of the hybrid Laplacian discretization shown in equation (2.55). Furthermore, the harmonic mean of the heat conductivities of the interacting particles is used, which results in the following expression

$$
\langle \nabla \cdot (\lambda \nabla T) \rangle_i = \sum_j \frac{2\lambda_i \lambda_j}{\lambda_i + \lambda_j} V_j \frac{2(T_j - T_i)}{r_{ij}^2} (\boldsymbol{x}_j - \boldsymbol{x}_i) \cdot \widetilde{\nabla}_i \widehat{w}_{ij},
\tag{3.81}
$$

with T_i and T_j being the temperatures and λ_i and λ_j the heat conductivities of the respective interacting particles i and j. The discretization of $\langle \boldsymbol{\tau} \rangle_i$ is material specific and therefore given in the respective section.

In addition, one has to stress, that a further discretization scheme based on the corrected second derivative of the kernel function, depicted in section 2.5, has been tested in the present work. However, worse results have been obtained at increased computational costs and the numerical results are not shown in the present work for that reason. For sake of completeness, the derivation of the used corrected second order derivative is presented in detail in the appendix E.

Discretization of the material balance

In the following paragraph, the material balance in the corrected and normalized SPH formalism is shown. Similar to the heat conduction term presented above, the molecular flux term in equation (3.10) is discretized using the corrected hybrid version of the second order derivative. So, the discretized form of the gradient of the molecular flux is expressed as

$$\left\langle \frac{\partial}{\partial x^\alpha} \left(b_k^M \, RT\varrho \frac{\partial \omega_k}{\partial x^\alpha} \right) \right\rangle_i = \sum_j \overline{b_{k,ij}^M \, RT_{ij} \, \varrho_{ij}} V_j \frac{2(\omega_{k,j} - \omega_{k,i})}{r_{ij}^2} (x_j - x_i) \cdot \widetilde{\nabla}_i \widehat{w}_{ij}, \quad (3.82)$$

where mobility of component k of particle i $b_{k,i}^M$ and particle j $b_{k,j}^M$ respectively, are concentrated in the harmonic mean of the mobility $\overline{b_{k,ij}^M \, RT_{ij} \, \varrho_{ij}}$, which is abbreviated as

$$\overline{b_{k,ij}^M \, RT_{ij} \, \varrho_{ij}} = \frac{2 \left(b_{k,i}^M \, RT_i \, \varrho_i \right) \left(b_{k,j}^M \, RT_j \, \varrho_j \right)}{\left(b_{k,i}^M \, RT_i \, \varrho_i \right) + \left(b_{k,j}^M \, RT_j \, \varrho_j \right)}. \quad (3.83)$$

The conditional equation for the mass fraction $\omega_{k,i}$ of component k carried by particle i results in

$$\varrho_i \frac{D\,\omega_{k,i}}{D\,t} = \sum_j \overline{b_{k,ij}^M \, RT_{ij} \, \varrho_{ij}} \, V_j \, \frac{2\left(\omega_{k,j} - \omega_{k,i}\right)}{r_{ij}^2} (x_j - x_i) \cdot \widetilde{\nabla}_i \widehat{w}_{ij} + MW_k r_k, \quad (3.84)$$

Treatment of boundaries

As stated in section 2.4, boundaries are the weak point in the conventional SPH framework. For non-viscous fluids, the zero pressure boundary condition at the free surface is approximately satisfied due to the particle deficit at the boundary [134, 155, 216]. As stated by Randles and Libersky, the missing particles act as if their pressure values are set to zero in the momentum equation, thereby approximately satisfying the free surface boundary condition. For problems involving material strength, this treatment results in an approximate stress-free boundary condition [155]. However, the picture changes, if a truly incompressible SPH scheme, as described in section 2.7.2, is used. Now, the zero pressure at the free surface needs to be specified as a boundary condition for solving the pressure Poisson equation. In addition, the implicit completion of the free surface boundary conditions is lost using corrected SPH approaches. Thus, the appropriate boundary condition has to be enforced on the surface. For this reason, the necessity for detecting free surface particles emerges. In the following paragraph, the used detection criteria are presented and the approach for enforcing the mathematical boundary condition is sketched. Subsequently, the used treatment of solid boundary conditions is exhibited.

Free surfaces

For the application of corrected as well as incompressible SPH approaches, a precise and reliable assessment of free boundary particles is crucial. As shown in section 2.7.2, a particle is regarded as being at the free surface, if its particle number density (or uncorrected density in general) obeys

$$n_i < \beta^S n^0, \qquad (3.85)$$

where the value of the parameter β^S is chosen in dependence of the smoothing length h [193, 217]. In the present work, the parameter β^S is varied between $\beta^S = 0.95$ for $h/L_0 = 1.05$ and $\beta^S = 0.85$ for $h/L_0 = 1.55$ with L_0 being the initial inter-particle spacing. However, in highly dynamic particle-based simulations, a different approach based on the work of Lee *et al.* is used, since the calculated particle density is dependent on the relative positions of its neighbor particles and so exposed to relative large fluctuations [178]. Furthermore, in the developed incompressible NCSPH approach, the incompressibility of particles at the free surface is not enforced explicitly, since a Dirichlet boundary condition is applied at the free surface when solving the pressure Poisson equation. For this reason, particle clustering cannot be excluded, leading to unwontedly high particle number densities at the free surface. In the incompressible NCSPH method, a different approach to identify free surface particles is deduced from equation (3.42). If the trace of the correction matrix A

$$tr(A) = \sum_j V_j(\boldsymbol{x}_j - \boldsymbol{x}_i) \cdot \nabla_i \widehat{w}_{ij} \qquad (3.86)$$

drops below a certain threshold, depended on the spatial dimension of the simulation, the regarded particle is considered as being on the free surface. For interior particles the result of the above equation is always equal to the dimension of the simulation. For example, for two-dimensional simulations, the result is equal to two for particles located in the bulk of the material. Particles are regarded to be free surface particles, if the result of equation (3.86) is lower than 1.5 for two-dimensional simulations. A similar approach for detecting free surface particles is used by Lee and coworkers, who suggest a free surface criterion of 2.4 for three-dimensional simulations [174, 178].

However, especially for highly dynamic flows, some bulk particles are still misleadingly regarded as free surface particles. To prevent this, an additional free surface criterion from Khayyer *et al.* is adopted for the incompressible NCSPH approach [172]. The criterion is based on the assumption, that the particle distribution in the neighborhood of free surface particles is non-symmetric in at least one of the spatial dimensions, while a more or less symmetric and spatially uniform particle neighborhood is encountered for bulk particles. Based on that, either one of the following criteria has to be fulfilled together with equation (3.86) or equation (3.85) for a particle to be considered as being on the free surface

$$\left| \sum_j r_{ij}^k \right| > L_0, \qquad \forall k = 1, ..., d \tag{3.87}$$

with d being the spatial dimension and L_0 the initial particle distance.

So far, a reliable method for detecting free surface particles has been assembled by using the combination of two criteria, given in equation (3.85) and (3.87) or equation (3.86) and (3.87), respectively. Therewith, the zero pressure boundary conditions can be accurately enforced on the free surface particles. In addition, the stress-free boundary condition has to be exerted. For this purpose, the deviatoric stress tensor $\boldsymbol{\tau}$ is rotated in the coordinate plane spanned by the outward pointing surface normal vector and the tangential vector tangential to the free surface. To apply the stress-free boundary condition, the normal component of the deviatoric stress tensor τ^{nn} and shear stress components τ^{nt} is set to

$$\tau_i^{nn} = 0 \qquad\qquad \tau_i^{nt} = 0 \tag{3.88}$$

Afterwards, the deviatoric stress tensor is rotated back into the Cartesian coordinate plane. The details of the coordinate transformation are shown in appendix D.

Wall boundaries

In contrast to the conventional SPH approach, the treatment of solid wall boundaries is simple within the framework of the corrected SPH method. As shown in figure 3.6, the wall needs to be modeled with a single layer of fixed boundary particles. To enforce the zero velocity slip boundary condition at the interface, the respective velocity $\boldsymbol{v} = 0$ is allocated to the boundary particle. The wall boundary particles are included in the pressure Poisson equation and are simply treated in the same way as ordinary incompressible fluid particles. Regarding the position of the wall, the fluid – wall interface is not located in between the fluid and wall boundary particles as in the conventional SPH approach, but right at the center of the wall boundary particles as depicted in figure 3.6.

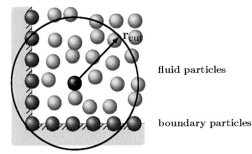

fluid particles

boundary particles

Figure 3.6: Sketch of the solid wall boundary treatment for the normalized corrected NCSPH approach. All boundary conditions are enforced on a single layer of boundary particles, with the position of the wall being colocated with the center of the boundary particles.

3.2.2.2 Normalized SPH model

In the section above, a rigorously corrected SPH formulation for multiphase systems with free surfaces has been presented. By using a renormalized smoothing function, the kernel interpolation is at least zero order consistent on the whole domain. In addition, by applying a gradient correction, the values of the gradient of linear functions are reproduced exactly on the whole domain. However, at the expense of increased numerical costs. Therefore, a SPH formalism with extenuated corrections, abbreviated as NSPH, is presented in the following section.

As in the NCSPH model shown above, the focus is laid on an accurate description of interfaces and free surface. Therefore, the reduced normalized SPH model, which is abbreviated by NSPH in the following, is based on the same renormalized kernel function (equation (3.36)), which is repeated below

$$\widehat{w}_{ij} = \frac{w_{ij}}{\sum_k V_k w_{ik}}. \tag{3.89}$$

Since constant functions are exactly interpolated on the whole domain, the kernel approximation near boundaries and interfaces is strongly improved. In contrast to the NCSPH model, the gradient operator is not obtained by differentiation of $\widehat{w}_{ij}$, but by differentiation of the conventional kernel, which is then corrected with the denominator of the Shepard kernel. This results in

$$\nabla w_{ij}^{NSPH} = \frac{\nabla w_{ij}}{\sum_k V_k w_{ik}}. \tag{3.90}$$

That approach has been introduced by Grenier et $al.$ and can be seen as a simplification of the complete derivative of the Shepard kernel with only keeping a constant correction of the gradient [119, 211]. The simplified gradient of the normalized kernel function ∇w_{ij}^{NSPH} is combined with the conventional first order derivative to give

$$\langle \nabla f \rangle_i = \sum_j V_j (f_j - f_i) \nabla w_{ij}^{NSPH} = \frac{1}{\sum_k V_k w_{ik}} \sum_j V_j (f_j - f_i) \nabla w_{ij}. \tag{3.91}$$

Despite the correction of the gradient, only constant functions are evaluated exactly. However, it has been shown by Grenier et $al.$ as well as Colagrossi and coworker, that the constant correction enhances the approximation at free surfaces and interfaces [119, 218]. In the following, the denominator is abbreviated by

$$\Gamma_i = \sum_k V_k w_{ik}, \tag{3.92}$$

with the summation being extended over all particles regardless of their nature. The phase-specific denominator is abbreviated by

$$\Gamma_i^{\Omega_i} = \sum_{k \in \Omega_i} V_k w_{ik}, \tag{3.93}$$

with the summation index being limited to particles of the same phase $k \in \Omega_i$. Before discussing the discretization schemes for the continuum balance equations, one has to stress, that the simplifications presented above were already introduced by Grenier and coworker in the context of weakly compressible flow [119, 211]. In the following paragraphs, the discretization method is transferred to the combined compressible and incompressible SPH approach developed in the work at hand.

Discretization of the continuity equation

Since both approaches, the Nsph and Ncsph approach, are based on the renormalized smoothing function, the same conditional equation for the density is used and repeated below

$$\varrho_i = \frac{\sum_{j\in\Omega_i} m_j w_{ij}}{\sum_{k\in\Omega_i} V_k w_{ik}}. \tag{3.94}$$

One has to emphasize, that the summations over the neighboring particles is limited to particles belonging to the same phase. The divergence term of the conditional equation for the particle volume is calculated using the first order derivative depicted in equation (3.91). This results in the following expression

$$\frac{D \ln(V_i/V_i^0)}{D\,t} = \frac{1}{\Gamma_i} \sum_j V_j (\boldsymbol{v}_j - \boldsymbol{v}_i) \cdot \nabla_i w_{ij}, \tag{3.95}$$

where the interactions between the particles are calculated over the whole particle neighborhood.

Discretization of the momentum balance

For the natural conservation of momentum, the gradient of the stress forces has to be derived from a variational principle as shown in section 2.6. The variational consistent gradient with respect to equation (3.91) derived by Grenier *et al.* reads as follows [211]

$$\langle \nabla f \rangle_i = \sum_j V_j \left(\frac{f_i}{\Gamma_i} + \frac{f_j}{\Gamma_j} \right) \nabla_i w_{ij}. \tag{3.96}$$

While the above equation conserves linear and angular momentum naturally, it is not even zeroth order consistent, since the gradient of a constant function does not vanish in boundary near regions. Nevertheless equation (3.96) is used to discretize the momentum balance within the Nsph framework. Again, a distinction between the compressible gas phase, where an absolute pressure value is obtained by a thermodynamic equation of state and incompressible fluid particles as well as solid particles has to be drawn. For the compressible particles $i \in \Omega_G$, the equation is expressed by

$$
\begin{aligned}
\varrho_i \frac{D\,\boldsymbol{v}_i}{D\,t} = &\sum_j V_j \left(\frac{\boldsymbol{\tau}_i}{\Gamma_i} + \frac{\boldsymbol{\tau}_j}{\Gamma_j} \right) \cdot \nabla_i w_{ij} - \sum_{j\in\Omega_G} V_j \left(\frac{p_i}{\Gamma_i} + \frac{p_j}{\Gamma_j} \right) \nabla_i w_{ij} \\
&- \sum_{j\notin\Omega_G} V_j \left(\frac{p_i}{\Gamma_i} + \frac{p_j^*}{\Gamma_j} \right) \nabla_i w_{ij} + \varrho_i \boldsymbol{f}_i \qquad \forall i \in \Omega_G.
\end{aligned} \tag{3.97}
$$

The scaled absolute pressure value of the solid and liquid particles p_j^* is calculated according to equation (3.51).

For the solid particles $i \in \Omega_S$, the momentum balance is modified to

$$\varrho_i \frac{D \boldsymbol{v}_i}{D t} = \sum_{j \in \Omega_S \cup \Omega_F} V_j \left(\frac{\boldsymbol{\sigma}_i}{\Gamma_i} + \frac{\boldsymbol{\sigma}_j}{\Gamma_j} \right) \cdot \nabla_i w_{ij} + \sum_{j \in \Omega_G} V_j \left(\frac{\tau_i}{\Gamma_i} + \frac{\tau_j}{\Gamma_j} \right) \cdot \nabla_i w_{ij}$$
$$- \frac{1}{\Gamma_i} \sum_{j \in \Omega_G} V_j p_j \nabla_i w_{ij} + \varrho_i \boldsymbol{f}_i \qquad \forall i \in \Omega_S. \tag{3.98}$$

As stated above, the volumetric strain or pressure of the solid particles is already included in the constitutive equation (3.52). In addition, the same interaction scheme for the solid particle pressure with the absolute pressure of the gaseous phase particles as described in section 3.2.1.1 is used.

For incompressible fluid particles, the momentum balance looks slightly different, since its solution is divided in two parts using the projection method. So, the momentum balance in the prediction step reads as

$$\varrho_i \frac{D \boldsymbol{v}_i}{D t} = \sum_{j \in \Omega_G \cup \Omega_F} V_j \left(\frac{\tau_i}{\Gamma_i} + \frac{\tau_j}{\Gamma_j} \right) \cdot \nabla_i w_{ij} + \sum_{j \in \Omega_S} V_j \left(\frac{\boldsymbol{\sigma}_i}{\Gamma_i} + \frac{\boldsymbol{\sigma}_j}{\Gamma_j} \right) \cdot \nabla_i w_{ij}$$
$$- \frac{1}{\Gamma_i} \sum_{j \in \Omega_G} V_j p_j \nabla_i w_{ij} + \varrho_i \boldsymbol{f}_i \qquad \forall i \in \Omega_F, \tag{3.99}$$

where interactions between incompressible particles are neglected in the calculation of the pressure gradient terms. The influence of the pressure of the neighboring, compressible phases Ω_G and Ω_S is included via the second and third term on the right hand side.

In contrast to solid and gaseous particles, where the pressure is obtained by an equation of state, the pressure of incompressible particles is calculated by solving a Poisson equation as depicted in equation (3.57). Using the conventional SPH gradient operator shown in equation (2.47), the divergence of the velocity field on the right hand side of the Poisson equation is written to

$$\langle \nabla \cdot \boldsymbol{v} \rangle_i = \sum_{j \in \Omega_F} V_j (\boldsymbol{v}_j - \boldsymbol{v}_i) \cdot \nabla_i w_{ij}, \tag{3.100}$$

where the summation is bounded to incompressible fluid particles only. Furthermore, the constant correction of the kernel is not taken into account in the latter equation due to the use of a simplified virtual particle method as shown below. For the discretization of the Poisson matrix, a modified Laplacian operator is derived. Based on equation (3.96), where the finite difference pressure gradient as shown in equation (3.61) is inserted, the following hybrid Laplacian discretization scheme evolves

$$\nabla \cdot \left(\frac{1}{\varrho_i} \nabla p_i \right) = \sum_j \frac{2 V_i V_j}{V_i + V_j} \left(\frac{1}{\varrho_i} + \frac{1}{\varrho_j} \right) (p_j - p_i) \frac{(\boldsymbol{x}_j - \boldsymbol{x}_i) \cdot \nabla_i w_{ij}}{r_{ij}^2}. \qquad (3.101)$$

One has to stress, that a harmonic mean of the particle volume is used to guarantee a symmetric form and that the constant gradient correction terms Γ_i and Γ_j are omitted. This is due to the used free surface boundary treatment approach as described later in section 3.2.2.2. Therewith, the Poisson equation results in

$$\sum_j a_{ij} p_j = b_i \qquad \forall i \in \Omega_F \qquad (3.102)$$

with the components of the symmetric positive definite matrix being

$$a_{ij} = \frac{2 V_i V_j}{V_i + V_j} \left(\frac{1}{\varrho_i} + \frac{1}{\varrho_j} \right) \frac{(\boldsymbol{x}_j - \boldsymbol{x}_i) \cdot \nabla_i w_{ij}}{r_{ij}^2} \quad \forall j \neq i,$$
$$a_{ii} = -\sum_{j \neq i} a_{ij}. \qquad (3.103)$$

The right hand side of equation (3.102) is discretized to

$$b_i = \sum_{j \in \Omega_F} V_j (\boldsymbol{v}_j^* - \boldsymbol{v}_i^*) \cdot \nabla_i w_{ij} \qquad (3.104)$$

for each incompressible particle i. As stressed in section 3.2.1 as well as 2.7.2, the asterik $*$ denotes the intermediate time after the predictor step. The coupling between incompressible and compressible phases is done in analogy to the NCSPH approach (see section 3.2.2.1) by applying a von Neumann boundary condition at the interface using the following gradient formula

$$\langle \nabla p \rangle_i = \sum_j V_j \left(\frac{p_j}{\Gamma_j} + \frac{p_i}{\Gamma_i} \right) \nabla_i w_{ij}. \qquad (3.105)$$

The same gradient discretization (equation (3.105)) is used to correct the particle velocity and position with the obtained pressure from the Poisson equation.

Discretization of the energy balance

In the paragraph below, the discretization of the energy balance shown in equation (3.6) is presented. Based on the simplifications introduced in section 3.2.2.1, the NSPH variant is expressed as

$$
\varrho_i c_{v,i} \frac{D\,T_i}{D\,t} = -\frac{p_i}{\Gamma_i} \sum_j V_j (\boldsymbol{v}_j - \boldsymbol{v}_i) \cdot \nabla_i w_{ij} - \langle \boldsymbol{\tau} \rangle_i : \sum_j V_j (\boldsymbol{v}_j - \boldsymbol{v}_i) \frac{\nabla_i w_{ij}}{\Gamma_i}
$$
$$
+ \sum_j \frac{2\lambda_i \lambda_j}{\lambda_i + \lambda_j} V_j \left(\frac{1}{\Gamma_j} + \frac{1}{\Gamma_i} \right) \frac{(T_j - T_i)}{r_{ij}^2} (\boldsymbol{x}_j - \boldsymbol{x}_i) \cdot \nabla_i w_{ij} \tag{3.106}
$$
$$
- \sum_k M W_k\, u_{k,i} \sum_h \nu_{hk}\, r_{h,i}.
$$

Again, for incompressible liquid particles as well as compressible solid particles, the first term on the right hand side is neglected. And in contrast to the Laplacian used for the discretization of the momentum equation, the constant correction of the kernel derivative Γ_i and Γ_j is applied in the discretization of the Laplacian operator of the heat conduction term despite the reflection of interior particles at the free surface using a virtual particle approach.

Discretization of the material balance

Below, the discretized version of the material balance based on the NSPH approximation is shown. Adopting the simplifications made in section 3.2.2.1 and using the hybrid version of the Laplacian operator, the material balance is expressed as

$$
\varrho_i \frac{D\,w_{k,i}}{D\,t} = \sum_j \overline{b_{k,ij}^M\, R\, T_{ij}\, \varrho_{ij}} V_j \left(\frac{1}{\Gamma_i} + \frac{1}{\Gamma_j} \right) \frac{(w_{k,j} - w_{k,i})}{r_{ij}^2} (\boldsymbol{x}_j - \boldsymbol{x}_i) \cdot \nabla_i w_{ij}
$$
$$
+ M W_k r_k, \tag{3.107}
$$

with the prefactor $\overline{b_{k,ij}^M\, R\, T_{ij}\, \varrho_{ij}}$ being defined in equation (3.83).

Treatment of boundaries

Within the NSPH approach, a distinction between fixed solid wall boundaries as well as free surfaces has to be made. First, the treatment of solid wall boundaries is sketched, before a simplified virtual particle approach for the free surface particles is described.

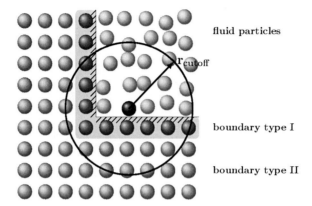

fluid particles

boundary type I

boundary type II

Figure 3.7: Solid wall boundary configuration with fluid particles shown in the vicinity of a fixed wall. Fixed wall boundary particles are classified in interior wall particles of type I and exterior wall boundary particles of type II within the NSPH framework.

Fixed wall boundaries

As stated in section 2.4, several approaches are used for modeling solid, impermeable boundaries. In the present work, we follow the solid boundary treatment introduced by Koshizuka *et al.* in his Moving Particle Semi-implicit method. That approach is also commonly used within the SPH framework [193, 217]. As shown in figure 3.7, the solid wall is modeled by fixed wall boundary particles of two types and the actual interface is located between the fluid and solid wall particles. Regardless of their type, the wall particles are spaced according to the initial particle configuration of the fluid. The thickness of the boundary particle layers is depended on the cutoff radius of the kernel function. For example, if using the cubic spline kernel, the depth of the boundary particle layer is at least two times the smoothing length h. The classification between the types of solid boundary particles is motivated by the different treatment of these particles for solving the Poisson equation of the projection algorithm. Only boundary particles of type I are included in solving the linear equation system. Hence, the fluid particles in the vicinity of the solid wall experience a repulsive force due to the pressure of the fixed wall. The additional lines of boundary particles of type II are added to keep the density of the boundary near fluid particles and boundary particles of type I to be consistent with the density of the bulk fluid. The density of the boundary particles of type II itself is not calculated but copied from the respective nearest neighbor boundary particle of type I. In addition, the pressure value of the nearest neighbor boundary particle of type I in the normal direction of the wall boundary is also copied to the respective boundary particle of type II. After this, no further distinction between the two types of boundary particles is made for all other calculations. For example, all boundary particles are included in the evaluation of the pressure gradient for the correction of the particle velocity and position based on equation (3.105). Besides, no distinction is made between the boundary particle types for calculating the viscous effects at solid walls, like the no velocity slip boundary condition following equation (3.97), (3.98) and (3.99). However, the velocity of the boundary particles has to be properly specified to ensure the zero velocity slip between the fluid and the wall. Details regarding the chosen approach are given in section 4.2 due to its strong dependence on the actual material behavior.

Free surfaces

While the treatment of fixed boundaries, e.g. solid walls, plays a negligible role in the context of this work, the treatment of free surfaces is crucial. In the conventional ISPH approach as shown in section 2.7.2, a Dirichlet boundary of zero pressure is applied at the free surface particles and the Poisson equation is not solved for these particles. So, the incompressibility of the fluid is not ensured for these particles. Even though, the free surface or interphase can be stress bearing in the current application. Therefore, a simplification of the *multiple boundary tangent method* introduced in section 2.4 is deployed to reduce the numerical expense for the simulation of open-porous materials [139, 140]. In the simplified approach, the particles are classified in free surface particles and bulk particles as sketched in figure 3.8. Free surface particles are detected using the condition shown in equation (3.85) and (3.87). The principle of the simplified approach is based on the point-reflection of neighboring bulk particles on the regarded free surface particle i. Whenever a boundary particle i is interacting with a bulk particle j, the bulk particle is point-reflected at the surface particle. The respective atmospheric pressure value is assigned to the reflected virtual particle j^* as it is shown in figure 3.8. The Dirichlet boundary condition at the free surface is specified as

$$p_{j^*} = 0. \tag{3.108}$$

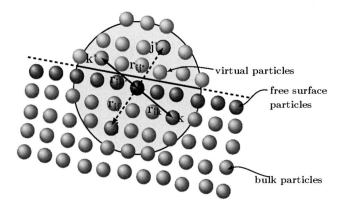

Figure 3.8: Simplified virtual boundary treatment at the free surface. Reconstruction of the interpolation domain by point–reflection of bulk particles at surface particle i.

Due to the point reflection, the following virtual particle properties hold:

$$r_{ij^*} = -r_{ij} \qquad (3.109\text{a})$$

$$\nabla_i w_{ij^*} = -\nabla_i w_{ij} \qquad (3.109\text{b})$$

Based on that, the Laplacian operator for free surface particles has to be modified for the free surface particle. To account for the influence of the virtual particle, the term on the principle diagonal of the coefficient matrix (equation (3.103)) has to be doubled to give

$$a_{ii} = -2 \sum_{j \neq i} \frac{2V_i V_j}{V_i + V_j} \left(\frac{1}{\varrho_i} + \frac{1}{\varrho_j} \right) \frac{(\boldsymbol{x}_j - \boldsymbol{x}_i) \cdot \nabla_i w_{ij}}{r_{ij}^2} \quad \forall\, j \in \text{bulk}. \qquad (3.110)$$

The non-diagonal elements a_{ij} of the regarded particle i stay unchanged as depicted in equation (3.103). This is also true for the right hand side of the Poisson equation (3.102), which is calculated for all fluid and boundary particles of type I according to equation (3.104). Thence, the Dirichlet boundary condition of zero pressure at the free surface is approximately ensured with the presented approach. Based on the obtained pressure profile, the velocity is corrected according to equation (3.105) for the bulk and surface particles within the NSPH approach. By using the presented virtual particle approach, which is a modification of the work of Ataie–Ashtiani and coworker, the incompressibility is also enforced for the free surface particles [194, 195, 219]. In contrast to the work of Ataie–Ashtiani *et al.*, the applied discretization scheme for the Laplacian operator is variational consistent with the divergence discretization scheme of the continuity equation and therewith, the right hand side of the Poisson equation (3.104).

For sake of completeness, one has to mention that for the conventional ISPH approach, the gradient term for free surface particles has to be also modified using the presented simplified virtual particle approach. The pressure of the virtual particle is set to

$$p_{j^\bullet} = -p_j. \tag{3.111}$$

For the surface particle i interacting with a bulk particle j, the pressure gradient is then simplified to

$$\langle \nabla p \rangle_i = 2 \sum_j V_j p_j \nabla_i w_{ij} \quad \forall\, j \in \text{bulk}, \tag{3.112}$$

with the pressure term of the surface particle being eliminated and the value of the gradient doubled at the free surface. One has to stress, that the latter approximation is not used in the NSPH approach, since the gradient (equation (3.105)) is already (partly) corrected using the constant denominator of the renormalized kernel for the kernel derivative as shown in equation (3.90).

To conclude, with the presented virtual particle approach, the consistency of the interpolation is approximately restored for free surface boundary particles. To reduce complexity in comparison to the multiple tangent boundary approach (see section 2.4), no special treatment is considered for boundary near particles. For these particles, the particle deficit remains. However, the most drastic effect of the particle deficit for particles at the free surfaces is eliminated. Consequently, the overall impact of the particle deficit is alleviated at least for relatively simple shaped boundaries. For complex boundary contours, especially non-symmetric surfaces, the accuracy of the interpolation is still deteriorated to a certain kind.

3.2.3   Simplified treatment of the gaseous phase – the hybrid approach

In the desired application, the reaction-induced formation of an open-porous system by release of a blowing agent, the actual dynamics of the gaseous blowing agent can be neglected, since the deformation of the compound matrix is mostly governed by the gaseous static pressure in the pores. However, the evolution of the pore volume and gas pressure has to be predicted over time. For this reason, a combined SPH–grid approach, the so called hybrid approach, is presented in the current paragraph. The hybrid approach is meant as a simplification in order to reduce complexity as well as numerical expense by neglecting the kinematics of the gaseous phase.

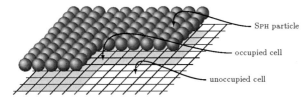

Figure 3.9: Schematic drawing of the unoccupied grid cells (□) and cells occupied by SPH particles (▨) in the hybrid SPH–grid approach.

In the hybrid approach, the fluid and solid phase is discretized using the NSPH or NCSPH approach as shown above, while the gaseous phase is discretized on a stationary grid. This concept is exemplary depicted in figure 3.9. The grid cell size has to be equal or smaller than the size of the SPH particles. The stationary grid is overlaid by the SPH particles and the grid cells are deactivated if superimposed by matrix particles or vice versa as outlined in figure 3.9. From the unoccupied grid cells, the pore volume can be calculated as it is sketched in figure 3.10. Thereby, it has to be ensured, that only connected cells are considered for the respective pore. This is done within a loop over all unoccupied grid cells, where all directly adjacent unoccupied grid cells are checked and marked in a list. The simple algorithm for the reconstruction of the pore volume is shown in appendix G. After its evaluation, every unoccupied cell is acquainted with all cells belonging to the same pore and the pore volume can be calculated. With known pore volume, the gas pressure can be estimated from a thermodynamic equation of state, since the mass of the decomposed wax equals the mass of the generated gas. Based on that, the force exerted by the pore pressure p_g on the fluid matrix surface can be computed in a similar way as with the *continuum surface force* (CSF) approach [204]. The force per unit volume acting on each SPH particle F_i, with i being the index of a SPH particle $i \in \Omega_p$, is simply calculated by

$$F_{i \in \Omega_p} = -p_g \boldsymbol{n}_i, \qquad (3.113)$$

with $\boldsymbol{n}_i$ being the normal vector of particle i to the matrix surface, whose absolute value represents the area to volume ratio of the interpolation region [204, 220]. For the hybrid approach, the estimation of the surface normal vector shown in equation (3.74) has to be modified and the summation is reduced to the SPH particles only and regardless of the particle type [205]. Therewith, the surface normal vector is simplified to

$$\boldsymbol{n}_i = \sum_{j \in \Omega_p} V_j \nabla_i w_{ij}. \qquad (3.114)$$

However, for the generation of an open-porous material, the latter equation is not suited as it will be shown in section 5.1.3 and the following approach is proposed

$$\boldsymbol{n}_i = \sum_{k \in \Omega_g} V_k \nabla_i w_{ik}. \qquad (3.115)$$

In the modified equation, the summation is limited to the unoccupied grid cells $k \in \Omega_g$, which comprise the pore as shown in figure 3.10.

One has to mention, that the conventional kernel function and its gradient have to be used in the NSPH as well as NCSPH method for estimating the normal vector. The theoretical background for converting a surface force F_S into a volumetric force F_V is given in detail in [204] and the originally equation, which is simplified to equation (3.113) is given by

$$F_V(\boldsymbol{x}) = F_S(\boldsymbol{x}) \hat{\boldsymbol{n}} \int (\chi(\boldsymbol{x}') - \chi(\boldsymbol{x})) \nabla w(\boldsymbol{x} - \boldsymbol{x}', h) d\boldsymbol{x}'. \qquad (3.116)$$

Furthermore, the modeling of the molecular transport in the multiphase system has to be modified. Since the underlying stationary grid is modeled by spatially fixed and constant volume cells, the used SPH discretization approach for the solid and fluid phase is simply extended to the gas phase modeled on the grid. In this way, no distinction

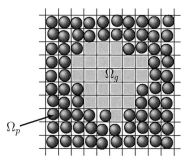

Figure 3.10: Schematic drawing of an closed pore Ω_g, represented by unoccupied cells
(▦) and bounded by SPH particles occupying the underlying grid cells Ω_p.

between the phases has to be made and all neighboring particles and unoccupied grid
cells are included in the summation of the SPH approximations.

As stressed above, the hybrid approach can be combined with the NCSPH as well as
NSPH discretization schemes. However, the right hand side of the pressure Poisson
equation of the correction step of the projection algorithm needs to be amended using the
NCSPH approach. In contrast to the purely SPH based approach, where a von Neumann
boundary condition is specified at the gas–liquid interface as shown in equation (3.70) to
(3.72), the interface has to be treated as a free surface since the gas phase is not resolved
by SPH particles. Hence, a Dirichlet boundary condition is specified at the liquid free
surface particles as mentioned in section 3.2.2.1. Though, the force vector F_i, as depicted
in equation (3.113), is acting on all boundary as well as boundary-near particles in
the prediction step of the projection scheme. So, these particles are accelerated and in
this way compressed. In the subsequent corrector step, the pressure of each particle
is calculated based on the divergence of the velocity field under the constraint of the
pressure Dirichlet boundary condition at the surface. Therewith, a pressure value of
zero is assigned to the free surface particles, while a higher pressure value is calculated
for the particles near the free surface. Based on that pressure field, the velocity of each
particle is corrected by evaluating the pressure gradient. And due to the higher pressure
of the free surface near particles, the expansion velocity of the free surface particles
caused by the force vector F_i is reduced by the contrary acting correction velocity. For

this reason, the overall velocity is diminished in an unphysical way. This behavior can be reduced in large parts by taking the velocity of the surrounding particles into account for calculating the divergence of the velocity field for the Poisson equation. Hence, the right hand side of the Poisson equation is modified to

$$b_i = \sum_{j \in \Omega_F} V_j (\boldsymbol{v}_j^* - \boldsymbol{v}_i^*) \cdot \widetilde{\nabla}_i \widehat{w}_{ij}^{\Omega_F} + \sum_{j \in \Omega_F} V_j (\Delta \boldsymbol{v}_j^* - \Delta \boldsymbol{v}_i^*) \cdot \widetilde{\nabla}_i \widehat{w}_{ij}^{\Omega_F}, \tag{3.117}$$

with $\Delta \boldsymbol{v}_i$ being the so-called XSPH velocity correction [49]. By taking into account the velocity of the neighboring particles to calculate a mean velocity

$$\Delta \boldsymbol{v}_i = \epsilon \sum_{j \in \Omega_F} \frac{2\, m_j}{\rho_j + \rho_i} (\boldsymbol{v}_j^* - \boldsymbol{v}_i^*) \widehat{w}_{ij}^{\Omega_F}, \tag{3.118}$$

particle inter-penetration is reduced. According to Monaghan, the parameter ϵ is commonly set to $\epsilon = 0.5$ [49]. In the current work, the parameter is reduced to $\epsilon = 0.25$. Further details on the *Extended* SPH or XSPH approach are given later in section 4.2.4. Equation (3.117) has been used by Colagrossi in the context of WCSPH and is applied in the present work within the hybrid – NCSPH framework only. The latter amendment is not needed for the hybrid – NSPH approach due to the different treatment of the free surface in the pressure Poisson equation as well as both purely SPH based approaches. In the subsequent section, the developed NCSPH as well as NSPH approach are compared to the state-of-the-art SPH approaches. Following this, the majority of the applied discretized governing equations for both approaches are compiled.

3.2.4 Comparison of the developed approaches with the state-of-the-art

In the following section, the developed SPH approaches are briefly compared and classed with the most prominent corrected SPH approaches available in the literature. At first, the NCSPH approach, as described in section 3.2.2.1, is compared to the corrected discretization schemes out of which the NCSPH approach has been developed. A special focus is laid on pointing out the differences between the approaches. Therefore, the NCSPH discretized governing equations are repeated in a simplified form, starting with the discretization scheme of the gradient, respectively divergence operator for the continuity equation as well as the stress tensor

$$\nabla \cdot f_i = \sum_j V_j \left(f_j - f_i \right) \widetilde{\nabla}_i \cdot \widehat{w}_{ij}, \qquad (3.119)$$

with $\widehat{w}_{ij}$ representing the renormalized kernel function and $\widetilde{\nabla}$ its corrected first order derivative. The divergence of the overall stress tensor $\boldsymbol{\sigma}$ in the momentum equation is calculated following

$$\nabla \cdot \boldsymbol{\sigma}_i = \sum_j V_j \left(\boldsymbol{\sigma}_j + \boldsymbol{\sigma}_i \right) \cdot \widetilde{\nabla}_i \widehat{w}_{ij}. \qquad (3.120)$$

As stated in section 3.2.2.1, the first discretization scheme can be seen as a combination of the Corrected Smoothed Particle Method (CSPM) of Chen *et al.* and the renormalized and corrected kernel function as derived by Bonet and Lok [87, 141]. The original CSPM first order derivative (equation (2.66)), which is based on a Taylor expansion of the SPH kernel estimate, is given as

$$\nabla \cdot f_i = \sum_j V_j \left(f_j - f_i \right) \widetilde{\nabla}_i \cdot w_{ij}. \qquad (3.121)$$

The difference to the NCSPH first order derivative (equation (3.119)) lies in the fact, that the conventional kernel function w and not a renormalized kernel $\widehat{w}_{ij}$ is used within the original CSPM approach.

Hence, the NCSPH discretization scheme also differs from the one proposed by Bonet and Lok. The latter authors suggest the following discretization scheme for the divergence of the stress tensor in the momentum equation [87]

$$\nabla \cdot \boldsymbol{\sigma}_i = \sum_j V_j \boldsymbol{\sigma}_j \cdot \widetilde{\nabla}_i \widehat{w}_{ij}. \qquad (3.122)$$

Even though, both approaches are first order consistent, the discretization of the pressure gradient term within the projection based SPH approach proved to be unstable if equation (3.122) is applied.

For the discretization of the momentum equation a further possibility has been developed by Vignjevic *et al.* [112], which is formulated as

$$\nabla \cdot \left(\frac{\boldsymbol{\sigma}_i}{\varrho_i} \right) = \sum_j m_j \left(\frac{\boldsymbol{\sigma}_j}{\varrho_j^2} + \frac{\boldsymbol{\sigma}_i}{\varrho_i^2} \right) \cdot \tilde{\nabla}_i \hat{w}_{ij}. \tag{3.123}$$

The divergence formula is also based on a renormalized kernel function and its corrected first order derivative. Yet, the latter equation is not variational consistent with the used continuity equation and therewith disregarded in the work at hand [87, 119].

So far, the presented corrected discretization schemes have mostly been applied within the weakly compressible SPH framework [87, 112, 136, 141–144, 221–223]. As an exception the work of Khayyer has to be mentioned. Khayyer *et al.* amended the conventional incompressible SPH approach with a corrected kernel approximation for the viscosity term [171, 172, 224]. One has to stress, that only the viscous term is corrected by Khayyer and coworkers with the purpose of preserving the angular momentum and not to increase the accuracy of the SPH approximation at the free surface. In contrast, Lind *et al.* used the CSPM first order derivative, as shown in equation (3.121), to discretize the pressure gradient in the correction step of the projection algorithm [225]. To the author's knowledge, no fully Corrected Normalized Incompressible SPH approaches comparable to the NCSPH approach are published in the literature. This is especially true, if considering the interaction of compressible and incompressible phases.

As stated in section 3.2.2.2, the foundations of the NSPH approach are based on the work of Grenier *et al.* [119, 211]. However, the renormalized SPH discretization is combined with the projection scheme based incompressible SPH approach in the present work and also extended to describe combined compressible as well as incompressible flow.

In the subsequent paragraph, the used discretized governing equation and constitutive relations using the NCSPH as well as NSPH approach are summarized.

3.2.5 Roundup of the NCSPH discretized equations

Governing equations differentiated by particle type $i \in \Omega_F, \Omega_G, \Omega_S$.

Constitutive equations

Overall stress tensor

$$\boldsymbol{\sigma}_i = \boldsymbol{\tau}_i - p_i \boldsymbol{I} \tag{3.124}$$

Deviatoric stress tensor (interactions with particle types $j \in \Omega_S \cup \Omega_G$ considered)

$$\begin{aligned}
\boldsymbol{\tau}_i^{\alpha\beta} = &\sum_{j \in \Omega_G \cup \Omega_F} \frac{2\eta_i \eta_j}{\eta_i + \eta_j} V_j \left((v_j^\alpha - v_i^\alpha)\widetilde{\nabla}_i^\beta \widehat{w}_{ij} + (v_j^\beta - v_i^\beta)\widetilde{\nabla}_i^\alpha \widehat{w}_{ij} \right) \\
&+ \sum_{j \in \Omega_S} \eta_i V_j \left((v_j^\alpha - v_i^\alpha)\widetilde{\nabla}_i^\beta \widehat{w}_{ij} + (v_j^\beta - v_i^\beta)\widetilde{\nabla}_i^\alpha \widehat{w}_{ij} \right) \\
&- \frac{2}{3}\delta^{\alpha,\beta} \sum_{j \in \Omega_G} \frac{2\eta_i \eta_j}{\eta_i + \eta_j} V_j (v_j^\gamma - v_i^\gamma)\widetilde{\nabla}_i^\gamma \widehat{w}_{ij}
\end{aligned} \tag{3.125}$$

Ideal gas equation of state

$$p_i = \frac{\varrho_i \, R \, T_i}{MW_i} \tag{3.126}$$

Figure 3.11: Discretized constitutive equations of compressible gaseous particles $i \in \Omega_G$ using the NCSPH approach.

Hook's law for isotropic elastic solids (interactions with particles $j \notin \Omega_S$ considered)

$$\begin{aligned}
\boldsymbol{\sigma}_i^{\alpha\beta} = &\int \left[2\mu_i \dot{\epsilon}_i^{\alpha\beta} + \lambda_i \delta^{\alpha\beta}\dot{\epsilon}_i^{\gamma\gamma} + \sigma_i^{\alpha\gamma}\Omega_i^{\gamma\beta} + \Omega_i^{\alpha\gamma}\sigma_i^{\gamma\beta} \right] Dt \\
&+ \sum_{j \notin \Omega_S} \eta_j \, V_j \left((v_j^\alpha - v_i^\alpha)\,\widetilde{\nabla}_i^\beta \widehat{w}_{ij} + (v_j^\beta - v_i^\beta)\widetilde{\nabla}_i^\alpha \widehat{w}_{ij} \right)
\end{aligned} \tag{3.127}$$

$$\dot{\epsilon}_i^{\alpha\beta} = \frac{1}{2} \sum_{j \in \Omega_S} V_j \left((v_j^\alpha - v_i^\alpha)\,\widetilde{\nabla}_i^\beta \widehat{w}_{ij} + (v_j^\beta - v_i^\beta)\widetilde{\nabla}_i^\alpha \widehat{w}_{ij} \right) \tag{3.128}$$

$$\Omega_i^{\alpha\beta} = \frac{1}{2} \sum_{j \in \Omega_S} V_j \left((v_j^\alpha - v_i^\alpha)\,\widetilde{\nabla}_i^\beta \widehat{w}_{ij} - (v_j^\beta - v_i^\beta)\widetilde{\nabla}_i^\alpha \widehat{w}_{ij} \right) \tag{3.129}$$

Figure 3.12: Discretized constitutive equations of compressible solid particles $i \in \Omega_S$ using the NCSPH approach.

overall stress tensor

$$\boldsymbol{\sigma}_i = \boldsymbol{\tau}_i - p_i \boldsymbol{I} \tag{3.130}$$

Newtonian fluid (interactions with particle types $j \in \Omega_S \cup \Omega_G$ considered)

$$\boldsymbol{\tau}_i^{\alpha\beta} = \sum_{j\in\Omega_F\cup\Omega_G} \frac{2\eta_i\eta_j}{\eta_i+\eta_j} V_j \left((\boldsymbol{v}_j^\alpha - \boldsymbol{v}_i^\alpha)\widetilde{\nabla}_i^\beta \widehat{w}_{ij} + (\boldsymbol{v}_j^\beta - \boldsymbol{v}_i^\beta)\widetilde{\nabla}_i^\alpha \widehat{w}_{ij} \right)$$
$$+ \sum_{j\in\Omega_S} \eta_i V_j \left((\boldsymbol{v}_j^\alpha - \boldsymbol{v}_i^\alpha)\widetilde{\nabla}_i^\beta \widehat{w}_{ij} + (\boldsymbol{v}_j^\beta - \boldsymbol{v}_i^\beta)\widetilde{\nabla}_i^\alpha \widehat{w}_{ij} \right) \tag{3.131}$$

Bingham fluid (interactions with particle types $j \in \Omega_S \cup \Omega_G$ considered)

$$\boldsymbol{\tau}_i^{\alpha\beta} = \left(\eta_i + \tau_o \frac{1-e^{-m\dot\gamma_i}}{\dot\gamma_i} \right) \sum_{j\in\Omega_F\cup\Omega_S} V_j \left((\boldsymbol{v}_j^\alpha - \boldsymbol{v}_i^\alpha)\widetilde{\nabla}_i^\beta \widehat{w}_{ij} + (\boldsymbol{v}_j^\beta - \boldsymbol{v}_i^\beta)\widetilde{\nabla}_i^\alpha \widehat{w}_{ij} \right)$$
$$+ \sum_{j\in\Omega_G} \frac{2\eta_i+\eta_j}{\eta_i+\eta_j} V_j \left((\boldsymbol{v}_j^\alpha - \boldsymbol{v}_i^\alpha)\widetilde{\nabla}_i^\beta \widehat{w}_{ij} + (\boldsymbol{v}_j^\beta - \boldsymbol{v}_i^\beta)\widetilde{\nabla}_i^\alpha \widehat{w}_{ij} \right) \tag{3.132}$$

$$\dot\gamma_i = \sqrt{\frac{1}{2}\sum_\alpha\sum_\beta \underline{\dot\epsilon}_i^{\alpha\beta}\,\underline{\dot\epsilon}_i^{\alpha\beta}} \tag{3.133}$$

$$\underline{\dot\epsilon}_i^{\alpha\beta} = \sum_{j\in\Omega_F} V_j \left((\boldsymbol{v}_j^\alpha - \boldsymbol{v}_i^\alpha)\widetilde{\nabla}_i^\beta \widehat{w}_{ij}^{\Omega_F} + (\boldsymbol{v}_j^\beta - \boldsymbol{v}_i^\beta)\widetilde{\nabla}_i^\alpha \widehat{w}_{ij}^{\Omega_F} \right) \tag{3.134}$$

Oldroyd–B fluid (no interactions with other particle types considered)

$$\boldsymbol{\tau}_i^{\alpha\beta} = \eta_i\frac{\lambda_{2,i}}{\lambda_{1,i}}\dot\epsilon_i^{\alpha\beta} + \boldsymbol{\tau}_{e,i}^{\alpha\beta} \tag{3.135}$$

$$\lambda_{1,i}\frac{D\boldsymbol{\tau}_{e,i}^{\alpha\beta}}{Dt} = \nabla^\gamma\boldsymbol{v}_i^\alpha\boldsymbol{\tau}_i^{\gamma\beta} + \nabla^\gamma\boldsymbol{v}_i^\beta\boldsymbol{\tau}_i^{\gamma\alpha} - \boldsymbol{\tau}_{e,i}^{\alpha\beta} + \eta_i\left(1-\frac{\lambda_{2,i}}{\lambda_{1,i}}\right)\dot\epsilon^{\alpha\beta} \tag{3.136}$$

$$\nabla^\gamma\boldsymbol{v}_i^\alpha = \sum_{j\in\Omega_F} V_j \left(\boldsymbol{v}_j^\alpha - \boldsymbol{v}_i^\alpha\right)\widetilde{\nabla}_i^\gamma\widehat{w}_{ij} \tag{3.137}$$

$$\dot\epsilon_i^{\alpha\beta} = \sum_{j\in\Omega_F} V_j \left((\boldsymbol{v}_j^\alpha - \boldsymbol{v}_i^\alpha)\widetilde{\nabla}_i^\beta \widehat{w}_{ij} + (\boldsymbol{v}_j^\beta - \boldsymbol{v}_i^\beta)\widetilde{\nabla}_i^\alpha \widehat{w}_{ij} \right) \tag{3.138}$$

Figure 3.13: Discretized constitutive equations for incompressible fluid particles $i \in \Omega_F$ using the NCSPH approach.

Balance equations

Continuity balance

$$\varrho_i = \frac{\sum_{j\in\Omega_G} m_j\, w_{ij}}{\sum_{k\in\Omega_G} V_k\, w_{ik}} \tag{3.139}$$

Volume balance

$$\frac{D\ln(V_i/V_i^0)}{Dt} = \sum_j V_j\,(\boldsymbol{v}_j - \boldsymbol{v}_i)\cdot\widetilde{\nabla}_i\widehat{w}_{ij} \tag{3.140}$$

Momentum balance

$$\varrho_i\frac{D\,\boldsymbol{v}_i}{D\,t} = \sum_j V_j(\boldsymbol{\tau}_i + \boldsymbol{\tau}_j)\cdot\widetilde{\nabla}_i\widehat{w}_{ij} - \sum_{j\in\Omega_G} V_j(p_i + p_j)\widetilde{\nabla}_i\widehat{w}_{ij}$$
$$- \sum_{j\in\Omega_F,\Omega_S} V_j(p_i + p_j^*)\widetilde{\nabla}_i\widehat{w}_{ij} + \varrho_i\boldsymbol{f}_i \tag{3.141}$$

Energy balance

$$\varrho_i c_{v,i}\frac{D\,T_i}{D\,t} = -p_i\sum_j V_j(\boldsymbol{v}_j - \boldsymbol{v}_i)\cdot\widetilde{\nabla}_i\widehat{w}_{ij} - \langle\boldsymbol{\tau}\rangle_i : \sum_j V_j(\boldsymbol{v}_j - \boldsymbol{v}_i)\widetilde{\nabla}_i\widehat{w}_{ij}$$
$$+ \sum_j \frac{2\lambda_i\lambda_j}{\lambda_i + \lambda_j}V_j\frac{2(T_j - T_i)}{r_{ij}^2}(\boldsymbol{x}_j - \boldsymbol{x}_i)\cdot\widetilde{\nabla}_i\widehat{w}_{ij}$$
$$- \sum_k MW_k\, u_{k,i}\sum_h \nu_{hk}\, r_{h,i} \tag{3.142}$$

Component balance

$$\varrho_i\frac{D\,\omega_{k,i}}{D\,t} = \sum_j \overline{b_{k,ij}^M\, R\, T_{ij}\,\varrho_{ij}}\, V_j\frac{2\,(\omega_{k,j} - \omega_{k,i})}{r_{ij}^2}(\boldsymbol{x}_j - \boldsymbol{x}_i)\cdot\widetilde{\nabla}_i\widehat{w}_{ij} + MW_k r_k \tag{3.143}$$

Figure 3.14: Discretized balance equations of compressible gaseous particles $i\in\Omega_G$ using the NCSPH approach.

Continuity balance

$$\varrho_i = \frac{\sum_{j \in \Omega_S} m_j \, w_{ij}}{\sum_{k \in \Omega_S} V_k \, w_{ik}} \tag{3.144}$$

Volume balance

$$\frac{D \ln(V_i/V_i^0)}{Dt} = \sum_j V_j \, (\boldsymbol{v}_j - \boldsymbol{v}_i) \cdot \widetilde{\nabla}_i \widehat{w}_{ij} \tag{3.145}$$

Momentum balance

$$\varrho_i \frac{D \boldsymbol{v}_i}{Dt} = \sum_{j \in \Omega_S \cup \Omega_F} V_j (\boldsymbol{\sigma}_i + \boldsymbol{\sigma}_j) \cdot \widetilde{\nabla}_i \widehat{w}_{ij} + \sum_{j \in \Omega_G} V_j (\boldsymbol{\tau}_i + \boldsymbol{\tau}_j) \cdot \widetilde{\nabla}_i \widehat{w}_{ij}$$
$$- \sum_{j \in \Omega_G} V_j p_j \widetilde{\nabla}_i \widehat{w}_{ij} + \varrho_i \boldsymbol{f}_i \tag{3.146}$$

Energy balance

$$\varrho_i c_{v,i} \frac{D T_i}{Dt} = \sum_j \frac{2 \lambda_i \lambda_j}{\lambda_i + \lambda_j} V_j \frac{2(T_j - T_i)}{r_{ij}^2} (\boldsymbol{x}_j - \boldsymbol{x}_i) \cdot \widetilde{\nabla}_i \widehat{w}_{ij}$$
$$- \langle \boldsymbol{\tau} \rangle_i : \sum_j V_j (\boldsymbol{v}_j - \boldsymbol{v}_i) \widetilde{\nabla}_i \widehat{w}_{ij} - \sum_k MW_k \, u_{k,i} \sum_h \nu_{hk} \, r_{h,i} \tag{3.147}$$

Component balance

$$\varrho_i \frac{D \omega_{k,i}}{Dt} = \sum_j \overline{b_{k,ij}^M \, R \, T_{ij} \, \varrho_{ij}} \, V_j \frac{2 \left(\omega_{k,j} - \omega_{k,i} \right)}{r_{ij}^2} (\boldsymbol{x}_j - \boldsymbol{x}_i) \cdot \widetilde{\nabla}_i \widehat{w}_{ij} + MW_k r_k \tag{3.148}$$

Figure 3.15: Discretized balance equations of compressible solid particles $i \in \Omega_S$ using the NCSPH approach.

Continuity balance

$$\varrho_i = \frac{\sum_{j \in \Omega_F} m_j \, w_{ij}}{\sum_{k \in \Omega_F} V_k \, w_{ik}} \tag{3.149}$$

Volume balance

$$\frac{D \ln(V_i/V_i^0)}{Dt} = \sum_j V_j \, (\boldsymbol{v}_j - \boldsymbol{v}_i) \cdot \widetilde{\nabla}_i \widehat{w}_{ij} \tag{3.150}$$

Momentum balance – predictor step

$$\varrho_i \frac{D \boldsymbol{v}_i}{Dt} = \sum_{j \in \Omega_G \cup \Omega_F} V_j (\boldsymbol{\tau}_i + \boldsymbol{\tau}_j) \cdot \widetilde{\nabla}_i \widehat{w}_{ij} + \sum_{j \in \Omega_S} V_j (\boldsymbol{\sigma}_i + \boldsymbol{\sigma}_j) \cdot \widetilde{\nabla}_i \widehat{w}_{ij}$$
$$- \sum_{j \in \Omega_G} V_j p_j \widetilde{\nabla}_i \widehat{w}_{ij} + \varrho_i \boldsymbol{f}_i \tag{3.151}$$

Momentum balance – corrector step

$$\sum_j a_{ij} p_j = \sum_{j \in \Omega_F} V_j (\boldsymbol{v}_j^* - \boldsymbol{v}_i^*) \cdot \widetilde{\nabla}_i \widehat{w}_{ij}^{\Omega_F} \quad \forall i \in \Omega_F \tag{3.152}$$

$$a_{ij} = V_j \frac{(\boldsymbol{x}_j - \boldsymbol{x}_i) \cdot (\frac{1}{\varrho_i} \widetilde{\nabla}_i \widehat{w}_{ij} - \frac{1}{\varrho_j} \widetilde{\nabla}_j \widehat{w}_{ij})}{r_{ij}^2} \quad \forall j \neq i \cup j \in \Omega_F \tag{3.153}$$

$$a_{ii} = - \sum_{j \neq i} a_{ij} \tag{3.154}$$

$$\nabla p_i = 2 \, p_i \widetilde{\nabla}_i \widehat{w}(r_{ii}) + \sum_{j \neq i \cup j \in \Omega_F} V_j (p_i + p_j) \widetilde{\nabla}_i \widehat{w}_{ij} \tag{3.155}$$

Energy balance

$$\varrho_i c_{v,i} \frac{D T_i}{Dt} = \sum_j \frac{2 \lambda_i \lambda_j}{\lambda_i + \lambda_j} V_j \frac{2(T_j - T_i)}{r_{ij}^2} (\boldsymbol{x}_j - \boldsymbol{x}_i) \cdot \widetilde{\nabla}_i \widehat{w}_{ij}$$
$$- \langle \boldsymbol{\tau} \rangle_i : \sum_j V_j (\boldsymbol{v}_j - \boldsymbol{v}_i) \widetilde{\nabla}_i \widehat{w}_{ij} - \sum_k MW_k \, u_{k,i} \sum_h \nu_{hk} \, r_{h,i} \tag{3.156}$$

Component balance

$$\varrho_i \frac{D \omega_{k,i}}{Dt} = \sum_j \overline{b_{k,ij}^M} \, R \, T_{ij} \, \varrho_{ij} \, V_j \frac{2 \left(\omega_{k,j} - \omega_{k,i} \right)}{r_{ij}^2} (\boldsymbol{x}_j - \boldsymbol{x}_i) \cdot \widetilde{\nabla}_i \widehat{w}_{ij} + MW_k r_k \tag{3.157}$$

Figure 3.16: Discretized balance of incompressible fluid particles $i \in \Omega_F$ using the NCSPH approach (without boundary/interface treatment as well as hybrid method amendments).

3.2.6 Roundup of the NSPH discretized equations

Governing equations differentiated by particle type $i \in \Omega_F, \Omega_G, \Omega_S$.

Constitutive equations

Overall stress tensor

$$\boldsymbol{\sigma}_i = \boldsymbol{\tau}_i - p_i \boldsymbol{I} \tag{3.158}$$

Deviatoric stress tensor (interactions with particle types $j \in \Omega_S \cup \Omega_G$ considered)

$$
\begin{aligned}
\tau_i^{\alpha\beta} &= \frac{1}{\Gamma_i} \sum_{j \in \Omega_G \cup \Omega_F} \frac{2\eta_i\eta_j}{\eta_i + \eta_j} V_j \left((\boldsymbol{v}_j^\alpha - \boldsymbol{v}_i^\alpha)\nabla_i^\beta w_{ij} + (\boldsymbol{v}_j^\beta - \boldsymbol{v}_i^\beta)\nabla_i^\alpha w_{ij} \right) \\
&+ \frac{1}{\Gamma_i} \sum_{j \in \Omega_S} \eta_i V_j \left((\boldsymbol{v}_j^\alpha - \boldsymbol{v}_i^\alpha)\nabla_i^\beta w_{ij} + (\boldsymbol{v}_j^\beta - \boldsymbol{v}_i^\beta)\nabla_i^\alpha w_{ij} \right) \\
&- \frac{2}{3}\delta^{\alpha,\beta} \frac{1}{\Gamma_i} \sum_{j \in \Omega_G} \frac{2\eta_i\eta_j}{\eta_i + \eta_j} V_j (\boldsymbol{v}_j^\gamma - \boldsymbol{v}_i^\gamma)\nabla_i^\gamma w_{ij}
\end{aligned}
\tag{3.159}
$$

Ideal gas equation of state

$$p_i = \frac{\varrho_i \, R \, T_i}{MW_i} \tag{3.160}$$

Figure 3.17: Discretized constitutive equations of compressible gaseous particles $i \in \Omega_G$ using the NSPH approach.

Hook's law for isotropic elastic solids (interactions with particle $j \notin \Omega_S$ considered)

$$
\begin{aligned}
\sigma_i^{\alpha\beta} &= \int \left[2\mu_i \dot{\epsilon}_i^{\alpha\beta} + \lambda_i \delta^{\alpha\beta} \dot{\epsilon}_i^{\gamma\gamma} + \sigma_i^{\alpha\gamma}\Omega_i^{\gamma\beta} + \Omega_i^{\alpha\gamma}\sigma_i^{\gamma\beta} \right] Dt \\
&+ \frac{1}{\Gamma_i} \sum_{j \notin \Omega_S} \eta_j \, V_j \left((\boldsymbol{v}_j^\alpha - \boldsymbol{v}_i^\alpha)\nabla_i^\beta w_{ij} + (\boldsymbol{v}_j^\beta - \boldsymbol{v}_i^\beta)\nabla_i^\alpha w_{ij} \right)
\end{aligned}
\tag{3.161}
$$

$$\dot{\epsilon}_i^{\alpha\beta} = \frac{1}{2\,\Gamma_i} \sum_{j \in \Omega_S} V_j \left((\boldsymbol{v}_j^\alpha - \boldsymbol{v}_i^\alpha)\nabla_i^\beta w_{ij} + (\boldsymbol{v}_j^\beta - \boldsymbol{v}_i^\beta)\nabla_i^\alpha w_{ij} \right) \tag{3.162}$$

$$\Omega_i^{\alpha\beta} = \frac{1}{2\,\Gamma_i} \sum_{j \in \Omega_S} V_j \left((\boldsymbol{v}_j^\alpha - \boldsymbol{v}_i^\alpha)\nabla_i^\beta w_{ij} - (\boldsymbol{v}_j^\beta - \boldsymbol{v}_i^\beta)\nabla_i^\alpha w_{ij} \right) \tag{3.163}$$

Figure 3.18: Discretized constitutive equations of compressible solid particles $i \in \Omega_S$ using the NSPH approach.

Overall stress tensor

$$\boldsymbol{\sigma}_i = \boldsymbol{\tau}_i - p_i \boldsymbol{I} \tag{3.164}$$

Newtonian fluid, (interactions with particle types $j \in \Omega_S \cup \Omega_G$ considered)

$$\boldsymbol{\tau}_i^{\alpha\beta} = \frac{1}{\Gamma_i} \sum_{j \in \Omega_F \cup \Omega_G} \frac{2\eta_i \eta_j}{\eta_i + \eta_j} V_j \left((\boldsymbol{v}_j^\alpha - \boldsymbol{v}_i^\alpha) \nabla_i^\beta w_{ij} + (\boldsymbol{v}_j^\beta - \boldsymbol{v}_i^\beta) \nabla_i^\alpha w_{ij} \right)$$
$$+ \frac{1}{\Gamma_i} \sum_{j \in \Omega_S} \eta_i V_j \left((\boldsymbol{v}_j^\alpha - \boldsymbol{v}_i^\alpha) \nabla_i^\beta w_{ij} + (\boldsymbol{v}_j^\beta - \boldsymbol{v}_i^\beta) \nabla_i^\alpha w_{ij} \right) \tag{3.165}$$

Bingham fluid, (interactions with particle types $j \in \Omega_S \cup \Omega_G$ considered)

$$\boldsymbol{\tau}_i^{\alpha\beta} = \left(\eta_i + \tau_o \frac{1 - e^{-m\dot{\gamma}_i}}{\dot{\gamma}_i} \right) \frac{1}{\Gamma_i} \sum_{j \in \Omega_F \cup \Omega_S} V_j \left((\boldsymbol{v}_j^\alpha - \boldsymbol{v}_i^\alpha) \nabla_i^\beta w_{ij} + (\boldsymbol{v}_j^\beta - \boldsymbol{v}_i^\beta) \nabla_i^\alpha w_{ij} \right)$$
$$+ \frac{1}{\Gamma_i} \sum_{j \in \Omega_G} \frac{2\eta_i \eta_j}{\eta_i + \eta_j} V_j \left((\boldsymbol{v}_j^\alpha - \boldsymbol{v}_i^\alpha) \nabla_i^\beta w_{ij} + (\boldsymbol{v}_j^\beta - \boldsymbol{v}_i^\beta) \nabla_i^\alpha w_{ij} \right) \tag{3.166}$$

$$\dot{\gamma}_i = \sqrt{\frac{1}{2} \sum_\alpha \sum_\beta \underline{\dot{\epsilon}}_i^{\alpha\beta} \underline{\dot{\epsilon}}_i^{\alpha\beta}} \tag{3.167}$$

$$\underline{\dot{\epsilon}}_i^{\alpha\beta} = \frac{1}{\Gamma_i^{\Omega_i}} \sum_{j \in \Omega_F} V_j \left((\boldsymbol{v}_j^\alpha - \boldsymbol{v}_i^\alpha) \nabla_i^\beta w_{ij} + (\boldsymbol{v}_j^\beta - \boldsymbol{v}_i^\beta) \nabla_i^\alpha w_{ij} \right) \tag{3.168}$$

Oldroyd–B fluid (no interactions with other particle types $j \notin \Omega_F$ considered)

$$\boldsymbol{\tau}_i^{\alpha\beta} = \eta_i \frac{\lambda_{2,i}}{\lambda_{1,i}} \dot{\epsilon}_i^{\alpha\beta} + \boldsymbol{\tau}_{e,i}^{\alpha\beta} \tag{3.169}$$

$$\lambda_{1,i} \frac{D \boldsymbol{\tau}_{e,i}^{\alpha\beta}}{D t} = \nabla^\gamma \boldsymbol{v}_i^\alpha \boldsymbol{\tau}_i^{\gamma\beta} + \nabla^\gamma \boldsymbol{v}_i^\beta \boldsymbol{\tau}_i^{\gamma\alpha} - \boldsymbol{\tau}_{e,i}^{\alpha\beta} + \eta_i \left(1 - \frac{\lambda_{2,i}}{\lambda_{1,i}} \right) \dot{\epsilon}^{\alpha\beta} \tag{3.170}$$

$$\nabla^\gamma \boldsymbol{v}_i^\alpha = \frac{1}{\Gamma_i} \sum_{j \in \Omega_F} V_j \left(\boldsymbol{v}_j^\alpha - \boldsymbol{v}_i^\alpha \right) \nabla_i^\gamma w_{ij} \tag{3.171}$$

$$\dot{\epsilon}_i^{\alpha\beta} = \frac{1}{\Gamma_i} \sum_{j \in \Omega_F} V_j \left((\boldsymbol{v}_j^\alpha - \boldsymbol{v}_i^\alpha) \nabla_i^\beta w_{ij} + (\boldsymbol{v}_j^\beta - \boldsymbol{v}_i^\beta) \nabla_i^\alpha w_{ij} \right) \tag{3.172}$$

Figure 3.19: Discretized constitutive equations for incompressible fluid particles $i \in \Omega_F$ using the NSPH approach.

Balance equations

Continuity balance

$$\varrho_i = \frac{\sum_{j\in\Omega_G} m_j w_{ij}}{\sum_{k\in\Omega_G} V_k w_{ik}} \tag{3.173}$$

Volume balance

$$\frac{D\ln(V_i/V_i^0)}{Dt} = \frac{1}{\Gamma_i}\sum_j V_j(\boldsymbol{v}_j - \boldsymbol{v}_i)\cdot\nabla_i w_{ij} \tag{3.174}$$

Momentum balance

$$\varrho_i\frac{D\boldsymbol{v}_i}{Dt} = \sum_j V_j\left(\frac{\tau_i}{\Gamma_i}+\frac{\tau_j}{\Gamma_j}\right)\cdot\nabla_i w_{ij} - \sum_{j\in\Omega_G} V_j\left(\frac{p_i}{\Gamma_i}+\frac{p_j}{\Gamma_j}\right)\nabla_i w_{ij} \tag{3.175}$$

$$-\sum_{j\notin\Omega_G} V_j\left(\frac{p_i}{\Gamma_i}+\frac{p_j^*}{\Gamma_j}\right)\nabla_i w_{ij} + \varrho_i\boldsymbol{f}_i \qquad i\in\Omega_G. \tag{3.176}$$

Energy balance

$$\varrho_i c_{v,i}\frac{DT_i}{Dt} = -\frac{p_i}{\Gamma_i}\sum_j V_j(\boldsymbol{v}_j-\boldsymbol{v}_i)\cdot\nabla_i w_{ij} - \langle\tau\rangle_i:\sum_j V_j(\boldsymbol{v}_j-\boldsymbol{v}_i)\frac{\nabla_i w_{ij}}{\Gamma_i}$$
$$\sum_j \frac{2\lambda_i\lambda_j}{\lambda_i+\lambda_j}V_j\left(\frac{1}{\Gamma_j}+\frac{1}{\Gamma_i}\right)\frac{(T_j-T_i)}{r_{ij}^2}(\boldsymbol{x}_j-\boldsymbol{x}_i)\cdot\nabla_i w_{ij} - \sum_k MW_k u_{k,i}\sum_h \nu_{hk}r_{h,i} \tag{3.177}$$

Component balance

$$\varrho_i\frac{Dw_{k,i}}{Dt} = \sum_j \overline{b_{k,ij}^M RT_{ij}\varrho_{ij}}V_j\left(\frac{1}{\Gamma_i}+\frac{1}{\Gamma_j}\right)\frac{(w_{k,j}-w_{k,i})}{r_{ij}^2}(\boldsymbol{x}_j-\boldsymbol{x}_i)\cdot\nabla_i w_{ij}$$
$$+ MW_k r_k \tag{3.178}$$

Figure 3.20: Discretized balance equations for compressible gas particles $i\in\Omega_G$ using the Nsph approach (without boundary/interface treatment amendments).

Continuity balance

$$\varrho_i = \frac{\sum_{j \in \Omega_S} m_j w_{ij}}{\sum_{k \in \Omega_S} V_k w_{ik}} \tag{3.179}$$

Volume balance

$$\frac{D \ln(V_i/V_i^0)}{D t} = \frac{1}{\Gamma_i} \sum_j V_j (\boldsymbol{v}_j - \boldsymbol{v}_i) \cdot \nabla_i w_{ij} \tag{3.180}$$

Momentum balance

$$\varrho_i \frac{D \boldsymbol{v}_i}{D t} = \sum_{j \in \Omega_S \cup \Omega_F} V_j \left(\frac{\boldsymbol{\sigma}_i}{\Gamma_i} + \frac{\boldsymbol{\sigma}_j}{\Gamma_j} \right) \cdot \nabla_i w_{ij} + \sum_{j \in \Omega_G} V_j \left(\frac{\boldsymbol{\tau}_i}{\Gamma_i} + \frac{\boldsymbol{\tau}_j}{\Gamma_j} \right) \cdot \nabla_i w_{ij} \tag{3.181}$$

$$- \frac{1}{\Gamma_i} \sum_{j \in \Omega_G} V_j p_j \nabla_i w_{ij} + \varrho_i \boldsymbol{f}_i \qquad i \in \Omega_S \tag{3.182}$$

Energy balance

$$\varrho_i c_{v,i} \frac{D T_i}{D t} = \sum_j \frac{2 \lambda_i \lambda_j}{\lambda_i + \lambda_j} V_j \left(\frac{1}{\Gamma_j} + \frac{1}{\Gamma_i} \right) \frac{(T_j - T_i)}{r_{ij}^2} (\boldsymbol{x}_j - \boldsymbol{x}_i) \cdot \nabla_i w_{ij}$$

$$- \langle \boldsymbol{\tau} \rangle_i : \sum_j V_j (\boldsymbol{v}_j - \boldsymbol{v}_i) \frac{\nabla_i w_{ij}}{\Gamma_i} - \sum_k MW_k u_{k,i} \sum_h \nu_{hk} r_{h,i} \tag{3.183}$$

Component balance

$$\varrho_i \frac{D w_{k,i}}{D t} = \sum_j \overline{b_{k,ij}^M R T_{ij} \varrho_{ij}} V_j \left(\frac{1}{\Gamma_i} + \frac{1}{\Gamma_j} \right) \frac{(w_{k,j} - w_{k,i})}{r_{ij}^2} (\boldsymbol{x}_j - \boldsymbol{x}_i) \cdot \nabla_i w_{ij}$$

$$+ MW_k r_k \tag{3.184}$$

Figure 3.21: Discretized balance equations for compressible solid particles $i \in \Omega_S$ using the Nsph approach.

Continuity balance

$$\varrho_i = \frac{\sum_{j \in \Omega_F} m_j w_{ij}}{\sum_{k \in \Omega_F} V_k w_{ik}} \qquad (3.185)$$

Volume balance

$$\frac{D \ln(V_i/V_i^0)}{D t} = \frac{1}{\Gamma_i} \sum_j V_j (\boldsymbol{v}_j - \boldsymbol{v}_i) \cdot \nabla_i w_{ij} \qquad (3.186)$$

Momentum balance – predictor step

$$\varrho_i \frac{D \boldsymbol{v}_i}{D t} = \sum_{j \in \Omega_G \cup \Omega_F} V_j \left(\frac{\boldsymbol{\tau}_i}{\Gamma_i} + \frac{\boldsymbol{\tau}_j}{\Gamma_j} \right) \cdot \nabla_i w_{ij} + \sum_{j \in \Omega_S} V_j \left(\frac{\boldsymbol{\sigma}_i}{\Gamma_i} + \frac{\boldsymbol{\sigma}_j}{\Gamma_j} \right) \cdot \nabla_i w_{ij}$$
$$- \frac{1}{\Gamma_i} \sum_{j \in \Omega_G} V_j p_j \nabla_i w_{ij} + \varrho_i \boldsymbol{f}_i \qquad i \in \Omega_F, \qquad (3.187)$$

Momentum balance – corrector step

$$\sum_j a_{ij} p_j = \sum_{j \in \Omega_F} (\boldsymbol{v}_j^* - \boldsymbol{v}_i^*) \cdot \nabla_i w_{ij} \quad \forall i \in \Omega_F \qquad (3.188)$$

$$a_{ij} = \frac{2 V_i V_j}{V_i + V_j} \left(\frac{1}{\varrho_i} + \frac{1}{\varrho_j} \right) \frac{(\boldsymbol{x}_j - \boldsymbol{x}_i) \cdot \nabla_i w_{ij}}{r_{ij}^2} \quad \forall j \neq i \cup j \in \Omega_F \qquad (3.189)$$

$$a_{ii} = - \sum_{j \neq i} a_{ij} \qquad (3.190)$$

$$\nabla p_i = \sum_{j \in \Omega_F} V_j \left(\frac{p_j}{\Gamma_j} + \frac{p_i}{\Gamma_i} \right) \nabla_i w_{ij} \qquad (3.191)$$

Energy balance

$$\varrho_i c_{v,i} \frac{D T_i}{D t} = \sum_j \frac{2 \lambda_i \lambda_j}{\lambda_i + \lambda_j} V_j \left(\frac{1}{\Gamma_j} + \frac{1}{\Gamma_i} \right) \frac{(T_j - T_i)}{r_{ij}^2} (\boldsymbol{x}_j - \boldsymbol{x}_i) \cdot \nabla_i w_{ij}$$
$$- \langle \boldsymbol{\tau} \rangle_i : \sum_j V_j (\boldsymbol{v}_j - \boldsymbol{v}_i) \frac{\nabla_i w_{ij}}{\Gamma_i} - \sum_k MW_k u_{k,i} \sum_h \nu_{hk} r_{h,i} \qquad (3.192)$$

Component balance

$$\varrho_i \frac{D w_{k,i}}{D t} = \sum_j \overline{b_{k,ij}^M R T_{ij} \varrho_{ij}} V_j \left(\frac{1}{\Gamma_i} + \frac{1}{\Gamma_j} \right) \frac{(w_{k,j} - w_{k,i})}{r_{ij}^2} (\boldsymbol{x}_j - \boldsymbol{x}_i) \cdot \nabla_i w_{ij}$$
$$+ MW_k r_k \qquad (3.193)$$

Figure 3.22: Discretized balance equations for incompressible fluid particles $i \in \Omega_F$ using the NSPH approach.

4

Validation and Verification

In the previous chapter 3, two numerical models for the simulation of structure formation processes have been proposed. In the following section, both models, namely the Normalized SPH (NSPH) and Normalized Corrected SPH (NCSPH) approach, are verified and validated by means of relatively simple and well defined test cases. First, the focus is laid on verifying the accuracy and stability of the basic projection scheme algorithm for single phase processes. Second, the applicability of the derived algorithms for predicting different material behavior is assessed. Last but not least, the application of the developed techniques for the quantitative description of multiphase flow is evaluated.

4.1 Validation and verification of the single phase algorithm

In the present section, the basic NSPH and NCSPH algorithm for modeling incompressible, single phase material is assessed. Thereby, a strong focus is laid on the validation, respectively verification by means of free surface problems. At first, the impact of the simplified virtual particle approach as well as the correction of the renormalized kernel gradient is investigated.

4.1.1 Heat conduction and treatment of free surface boundaries

As shown in section 3.1, the second order derivative occurs in several physical models used in the field of structure generation. For example, structure formation processes are in general governed by the process temperature as well as the composition of the surrounding gas phase. Hence, the energy balance as well as material balance has to be

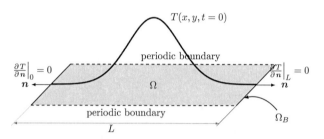

Figure 4.1: Computational domain of the two-dimensional slab with initial temperature profile $T(x, y, t = 0)$.

solved. Moreover, the discretization of the Laplacian operator plays a crucial role in projection based incompressible SPH algorithms as presented in section 2.7.2. In this context, the accurate evaluation of the second order derivative is crucial for most of the underlying physical processes in the regarded application. In the following section, the accuracy of the applied Laplacian operators are analyzed with the help of a simple heat transfer test case. In this way, implications of the boundary treatment as well as correction approaches are discussed. In addition, the influence of the smoothing length on the accuracy of the simulation is surveyed and a shortened convergence study is performed.

In the following, the heat conduction in a rectangular slab with an initially Gaussian distributed temperature field in horizontal direction is considered. Due to the relative simplicity of the test case geometry and the static particle setup, a best case study of the accuracy of the Laplacien operators and boundary models is made, since the accuracy is not deteriorated by particle disorder. The respective computational domain is shown in figure 4.1, together with the initial temperature field $T(x, y, t = 0)$. In horizontal directions, homogeneous von Neumann boundary conditions are applied, while periodic boundary conditions are deployed at the upper and lower boundaries, which are in this way considered adiabatic. Therewith, the two-dimensional problem is reduced to a quasi-one-dimensional one. The thermal diffusivity of the material is set to $\alpha = \frac{\lambda}{\varrho c_p} = 0.1 \, \frac{ul^2}{s}$ for all particles. The initial temperature distribution is given by

$$T(x, y, t = 0) = T_{max} e^{0.03(x-L/2)^2}. \tag{4.1}$$

The slab is resolved with 33 particles in horizontal direction. In the vertical direction, the slab is resolved by ten layers of particles, which is considered as sufficient due to the periodic boundary condition. The cubic spline kernel, as shown in equation (2.34), is used as smoothing function and the cutoff radius is at first chosen to be 3.1 times the initial particle spacing L_0. This results in a smoothing length to initial particle spacing ratio of $h/L_0 = 1.55$, which is close to the considered optimal choice given in the literature with $h = 1.6$ and explained in section 2.1.4 [55, 61, 76]. The parameter of the computational domain are summarized in table 4.1 together with further computational details.

For comparison, the heat equation is solved on a one dimensional grid, using a second order central finite difference scheme. For time integration, a fourth order Runge-Kutta method is used. In the following, the finite difference solution is abbreviated by FDM. Therewith, the local deviation in percent of the SPH solutions from the FDM solution is calculated. Furthermore, the average deviation in percent $\bar{\epsilon}$ from the finite difference solution is calculated as

$$\bar{\epsilon} = \frac{100}{L} \int_0^L \frac{T(\boldsymbol{x}) - T^{\mathrm{FDM}}(\boldsymbol{x})}{T^{\mathrm{FDM}}(\boldsymbol{x})} d\boldsymbol{x}. \tag{4.2}$$

As already stressed, the accurate treatment of free surfaces plays an important role for the regarded applications. Therefore, the implications of the simplified virtual particle approach, presented in section 3.2.2.2, are compared to a rigorous treatment as shown in section 2.4. Moreover, the influence of a truncated boundary on the simulation results is discussed, since it is common in the SPH literature to neglect any special boundary treatment at free surfaces for ease of simplicity [49, 67, 119]. Thus, the conventional SPH Laplacian operator (equation (2.55)) is used without any special boundary treatment for the simulation of the heat conduction test case and compared to the results obtained by the so called rigorous virtual particle approach. For both methods, the former being abbreviated by SPH and the latter by SPH RVP, the temporal evolution of the dimensionless temperature profiles over the slab length is shown in figure 4.2 and compared to the FDM solution, which is depicted by the solid line. In the lower part of the figure, the local deviation of the SPH approaches from the FDM solution is resolved. The particle temperature obtained without boundary treatment (SPH) are marked by $\bigcirc$ on the right half of figure 4.2 and the profiles are given at the time steps $0\,s, 25\,s$ and $100\,s$. On the left half of figure 4.2 the solution of the rigorous virtual particle approach using the conventional SPH method (SPH RVP) is shown at the same instants of time. Using the rigorous virtual particle approach, all particles, whose interpolation domain are truncated by a boundary, are mirror-imaged at the interface. Thereby, the whole interpolation domain is reconstructed in the SPH RVP version. The

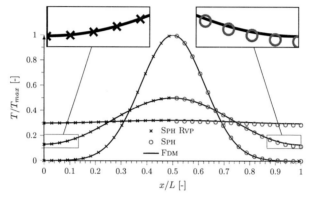

temperature evolution along slab length

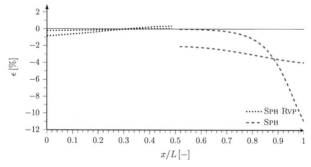

local deviation of both SPH solutions from FDM solution

Figure 4.2: Heat conduction with free boundary: Comparison of the conventional SPH method with the rigorous virtual particle boundary treatment (SPH RVP, left) with the conventional SPH method without any special boundary treatment (SPH, right).

respective particle temperatures are marked with ×. As one can see, the conventional SPH approach in conjunction with the rigorous virtual particle approach (SPH RVP) is in good agreement with the FDM solution. In contrast, the approach without boundary treatment (SPH) shows a rather large discrepancy to the FDM solution at all times. Regarding the deviation from the FDM solution shown in the lower figure, the largest discrepancy occurs at the free surface for both methods. However, the absolute value of the deviation with approximately 11 % after 25 s is much larger in the case of the SPH approach without special boundary treatment. Over time, the error profiles flatten out in both cases, with the deviation being distributed evenly over the whole spatial domain. The average deviation of the temperature profile at the steady state lies in the area of 3 % without boundary treatment. In contrast, by using the rigorous virtual boundary particle approach, the maximum deviation is reduced to less than 1 % after 25 seconds and approximately 0.6 % at the steady state. As expected, the rigorous virtual particle approach eliminates the particle deficit at the free boundary and restores the consistency of the interpolation. Admittedly, at a higher numerical expense and complexity as discussed in section 2.4.

Thence, the impact of the simplified virtual particle approach on the accuracy of the numerical solution is investigated together with a renormalization of the kernel function as introduced in the NSPH approach of section 3.2.2.2. Due to the renormalized kernel function, the density is not decreasing towards the free surface. For the discretization of the Laplacian operator, the version shown in equation (3.106) is used. As depicted in figure 4.3, the temperature profiles of the NSPH approach are compared to the ones obtained using the conventional SPH formalism with the rigorous virtual particle approach. The particle temperatures obtained by the NSPH approach in conjunction with the simplified virtual particle boundary treatment are shown on the right, marked by +, and are in good agreement with the FDM solution. The largest deviation with about 1.5 % occurs not directly at the boundary, but in the second particle row. This is obvious, since the particle deficit is only corrected for particles directly at the boundary and not for particles in the following rows. Even so, the error is relatively small, with the maximum average deviation from the FDM solution being calculated to less than 1 % over the whole simulation time. Compared to the deviations of the rigorous virtual particle approach, this is only a marginal deterioration of the accuracy under the presetting of a static particle arrangement.

As stressed above, the presented test case can be considered as a best case estimation due to the regular particle setup and geometry of the free surface. Thus, inferior results

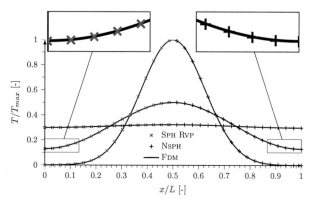

temperature evolution along slab length

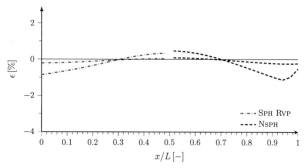

local deviation of the SPH solutions from FDM solution

Figure 4.3: Heat conduction with free boundary: Comparison of the conventional SPH method with the rigorous virtual particle boundary treatment (left, SPH RVP) with the renormalized NSPH with the simplified virtual particle boundary treatment (right, NSPH).

have to be expected for more complex boundary geometries. Therefore, the impact of the Normalized Corrected SPH discretization scheme on the solution of the simple static test case at hand is investigated. For discretizing the second order spatial derivative of the temperature field, the hybrid formulation of the Laplacian operator shown in equation (3.81) is used. As stressed in section 3.2.2.1 in detail, the renormalized kernel is differentiated and corrected. So, the boundary does not need to be reconstructed,

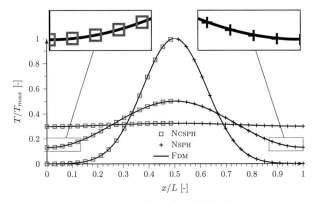

temperature evolution along slab length

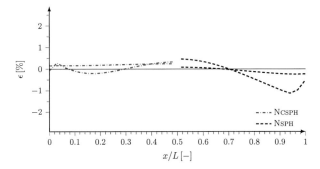

local deviation of both SPH solutions from FDM solution

Figure 4.4: Heat conduction with free boundary: The Normalized Corrected SPH method (NCSPH, left) in comparison to the Renormalized SPH with the simplified virtual particle boundary treatment (NSPH, right).

eliminating the occurrence of possible boundary effects due to incomplete reconstruction of the interpolation domain. Needless to say, the mathematical boundary conditions still have to be applied at the boundary particles. In this way, the homogeneous von Neumann boundary condition is applied for particles at the free surface. This is done by calculating the gradient of the temperature field, projecting it onto the unit normal vector and subtracting the resulting value from the right hand side of equation (3.81) of the respective boundary particle. In figure 4.4, the temperature profile obtained with the NCSPH formalism is compared to the FDM solution and the renormalized NSPH approach with the simplified virtual boundary treatment at the time steps $0\,s$, $25\,s$ and $100\,s$. The NCSPH solution, depicted by □, is in excellent agreement with the FDM solution. As one can see, the local deviation in percent of the NCSPH approach from the FDM solution is considerably smaller as compared to the respective conventional SPH solution. The maximal average deviation is calculated to less than $0.2\,\%$ over the whole simulation time.

So far, it has been demonstrated, that the accuracy of the applied discretization methods for the second order spatial derivative is strongly affected by the chosen boundary treatment. Thereby, the impact of a truncated interpolation domain has been demonstrated together with possible approaches to approximately reconstruct the interpolation domain and enforce the desired boundary condition. Furthermore, it has been demonstrated, that the simplified virtual particle approach, with its approximate reconstruction of the interpolation domain, leads to sufficiently accurate results in the present test case with a static particle configuration. Since more complex boundary geometries are expected during the morphogenesis of open-porous materials, a rigorous virtual particle approach, like the multiple boundary tangent method presented in section 2.4, as well as the Normalized Corrected SPH approach have been investigated. Both approaches proved to be very accurate, while the NCSPH approach seems to be better suited for the quantitative simulation of open-porous materials due to its implicit consistency at free surfaces.

In the following paragraph, the average deviation of the SPH approaches from the FDM solution is discussed as a function of the kernel cutoff radius r_{cut}. As stated above, the optimal smoothing length is estimated to approximately $h = 1.6\,L_0$, with L_0 being the initial particle spacing in the literature. For the cubic spline kernel, this results in a cutoff radius of $r_{cut} = 3.2\,L_0$. However, the optimal cutoff radius is determined for bulk simulations only, where the influence of a truncated interpolation domain is neglected. In order to study the impact of a free boundary, the cutoff radius r_{cut} is increased from 1.5 over 2.1 and 3.1 to 4.1 times the initial particle distance L_0. The particles are still arranged on a regular lattice. Besides the methods already introduced above, the investigation is expanded with the conventional SPH approach in conjunction with the simplified virtual particle approach. That combination is abbreviated by

SPH VP. In return, the conventional SPH approach without boundary treatment is disregarded in the following. In figure 4.5, the deviations of the respective methods are dawn in dependence of the cutoff radii at the time step $t = 25\,s$. As one can see, all methods possess approximately the same order of magnitude in accuracy, if the smallest cutoff radius of $r_{cut} = 1.5\,L_0$ is neglected. For the lowest cutoff radius of $r_{cut} = 1.5\,L_0$, the inaccuracy of the interpolation due to the small support domain can be reduced by correction of the kernel gradient, as done in the NCSPH approach. However, in actual simulations, where particle movement is considered, such a small interpolation domain will most likely lead to numerical instabilities. For a cutoff radius of $r_{cut} = 2.1\,L_0$, all methods show a similar accuracy, with the conventional, uncorrected SPH approaches being slightly more accurate. By increasing the cutoff radius towards

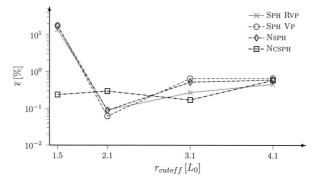

Figure 4.5: Average deviation $\bar{\epsilon}$ over the kernel cutoff radius.

$r_{cut} = 3.1\,L_0$ slightly worse results are obtained for the uncorrected SPH approaches, while the deviation of the NCSPH approach from the FDM solution is slightly reduced. This may indicate that the influence of the truncated boundary is effectively reduced by the kernel renormalization and correction of the gradient. The again increasing error with increasing cutoff radius of $r_{cut} = 4.1\,L_0$ might be explained by the smoothing of the temperature over a larger domain. Even though only minor differences are observed between the methods for cutoff radii larger than 1.5, the highest accuracy is obtained for a cutoff radius of $2.1\,L_0$. Therefore, the optimal cutoff radius estimated in the literature to $r_{cut} = 3.2\,L_0$ as stated above is reduced in the presence of a truncated boundary.

Parameter		Value
maximum time step	Δt	$10^{-4}\,s$
initial particle distance	L_0	$0.025\,ul$
slab width	L	$33\,L_0$
slab height	L	$10\,L_0$
cutoff radius	r_{cut}	$1.5 - 4.1\,L_0$
surface bound	β^S	0.8
thermal diffusivity	α	$0.1\,\frac{ul^2}{s}$

Table 4.1: Computational details of the heat conduction test case. All measures of length are given in chosen units abbreviated by ul as unit length.

For sake of completeness, one has to mention, that the heat conduction test case has been also investigated using a discretization formula based on the corrected second order derivative of the kernel function, as shown in appendix E. However, inferior results have been obtained in comparison to the upper approaches at higher computational costs. Hence, the results are not reported in the present work. Supplementary, the impact of higher order spline kernels has been explored. For the investigated M_6 and M_7 spline kernels as presented in section 2.1.4, no significant increase in accuracy has been observed. In contrast, slightly worse results have been obtained, since the interpolation domain has to be increased for higher order kernels. Thereby, the interpolation domain of more particles is truncated by the boundary.

In the present section, the influence of different boundary treatment techniques has been investigated together with a parametric study of the influence of the cutoff radius on the accuracy of the SPH approximation. Yet, the investigation was based on a fixed particle setup. In the following section, the accuracy and stability of the developed incompressible NSPH and NCSPH approaches is checked for freely moving particles in an almost static configuration.

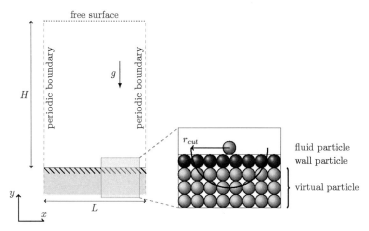

Figure 4.6: Computational domain of the two–dimensional channel with a free surface and detail of the solid boundary representation.

4.1.2 Hydrostatic pressure in a channel with a free surface

In the previous section, the accuracy of various SPH approximations as well as boundary reconstruction techniques has been examined with a spatially fixed particle setup. In the following section, the accuracy of both proposed symmetric Laplacian operators, shown in equation (3.62) and (3.101), is examined and the stability of the developed incompressible SPH approaches is investigated by means of a static two-dimensional free surface test case. At first sight, the test case seems trivial, but at low or negligible particle dynamics, the treatment of the isotropic pressure term as well as its gradient is crucial for the stability of the Smoothed Particle simulation scheme. Errors occurring at the free surface due to the particle deficit can grow over time and propagate towards the inside of the fluid domain. Since particle simulations are mostly applied to highly dynamic systems, the stability of the numerical scheme at low particle dynamics is insufficiently investigated in the literature as exemplary stated by Oger and coworker [124, 173, 219, 226]. Accordingly, the conventional Incompressible SPH method, as described in section 2.7.2, the incompressible Normalized SPH and the incompressible Normalized Corrected SPH approach are applied to the following test case and the results are compared to a weakly compressible SPH approach (WCSPH) in addition. These previously mentioned difficulties are even more prominent using the WCSPH approach, where the particle pressure is calculated by an equation of state as sketched in section 2.7.1. Due to the weak, but immanent compressibility of the particles, particles get disorganized during the course of the simulation, leading to spurious oscillation of the pressure field. Even though, the fluid is under compression in the presented test case, some particles may experience tensile forces due to negative pressure values of neighboring particles. Therewith, particles attract each other via the pressure gradient term, possibly triggering the tension instability phenomena as e.g. reported by Issa for an weakly compressible fluid [46, 74, 75, 173, 213]. These oscillations and their consequences can be significantly reduced by inserting artificial viscosity and artificial pressure terms into the momentum equation as reported by many authors [49, 55, 57, 78, 173, 213, 227]. However, these terms have to be parameterized, and the parameters have to be chosen carefully in order to not affect the original solution. Moreover, the occurrence of chaotic pressure values and velocity distributions is reported by Fang *et al.* if badly parameterized artificial forces are used [228]. In order to meliorate that problem, time-dependent artificial parameters have been introduced, which adjust the impact of the artificial stabilizing force depending on the local flow field [229, 230]. Although, strong tensile instabilities are not expected using the projection based approach for modeling incompressible flow, errors at the free surface can still be introduced via the free surface boundary condition for the pressure Poisson equation and the gradient of the pressure field, by which the particle velocities and positions are corrected. These errors may also grow over time and propagate towards the inside of the fluid domain. Consequently, a two-dimensional free surface channel is set up as depicted in figure 4.6. Periodic boundary conditions are applied in horizontal directions. The computational domain is discretized with 200 fluid particles, distributed on a 10×20 regular lattice. The fluid is treated as non-viscous. The solid wall boundary is represented by one layer of ten fixed particles (type I). For the conventional ISPH and incompressible NSPH approach, additional three layers of fixed wall boundary particles of type II are included in the simulation as described in section 3.2.2.2, since the total thickness of the boundary layer needs to be at least equal to the extent of the kernel support. For the conventional ISPH approach as well as the incompressible NSPH approach, the free surface boundary condition as described in section 3.2.2.2 is applied to the boundary particle at the free surface. The free surface boundary treatment approach used within the NCSPH framework is described in section 3.2.2.1. For comparison, a weakly compressible SPH solution is presented in the following. The WCSPH algorithm is based on the NSPH approach shown in section 3.2.2.2, with the projection step being replaced by a Tait's equation of state shown in equation (2.84) and rewritten in the following for sake of completeness

$$p(\varrho) = \frac{\varrho_0 c_0^2}{\gamma} \left[\left(\frac{\varrho}{\varrho_0} \right)^\gamma - 1 \right].$$ (4.3)

The parameters ϱ_0, c_0 and γ are the incompressible density, nominal speed of sound and the polytropic constant, respectively. To resemble the behavior of a weakly compressible medium, the polytropic constant of water $\gamma = 7$ is used. All further computational details and parameter values are given in table 4.2.

In the following paragraph, the results of the regarded test case are presented. For the simulation, a reference density ϱ^0 is applied to all particles as initial condition. Likewise, all particles are at rest and under the influence of gravity. After a settling phase, a constant pressure profile should emerge. The obtained pressure profiles over the channel height H after $t = 0.001\,s$ (100 time steps) are depicted in figure 4.7 on the left, together with the analytical solutions (solid line) as a reference. For sake of clarity, the SPH solutions and the respective analytical solution are shifted in vertical direction by $20\,Pa$ each. Besides, the absolute error Δp of the SPH approaches with respect to the analytical solution is displayed in the right diagram. As one can see, the WCSPH method (∇) is not capable of reproducing the analytical solution. This behavior is not surprising, since no additional tensile stability control techniques, such as the concept of artificial viscosity, are used [49, 213]. In order to obtain a stable and accurate WCSPH model, these approaches can also be supplemented by a density reinitialization scheme as it has been done by Colagrossi and Landrini [81]. However, the former approaches are dependent on parameters, which have to be adjusted according to the actual particle dynamics, as mentioned above. Therefore, the WCSPH approach is not suitable for the predictive simulation of structure forming processes. In contrast, the analytical pressure profile is sufficiently resembled by the SPH schemes based on the projection approach. The most accurate results are obtained using the Normalized Corrected SPH scheme (o), with the absolute error being smaller than $0.1\,Pa$ as depicted on the right side of figure 4.7. Therewith, the newly presented symmetric Laplacian operator (equation (3.62)) is perfectly suited for the discretization of the pressure Poisson equation.

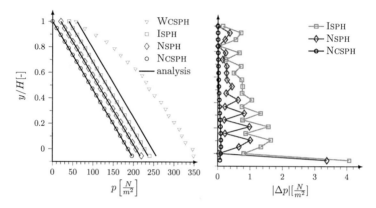

Figure 4.7: Left: Pressure profile after $t = 0.001\,s$ of the incompressible NCSPH (o), incompressible NSPH ($\diamond$), ISPH ($\square$) and WCSPH ($\triangledown$) approach over the channel height H in comparison with the analytical solution (solid line). Each pair of results is shifted on the abscissa for ease of display; Right: absolute error of the corresponding method with respect to the analytical solution.

Regarding the NSPH approach ($\diamond$), the absolute error increases in general in comparison to the NCSPH result. The error shows a sawtooth like profile over the channel height and is especially pronounced in the vicinity of the solid wall. There, the absolute error value adds up to $\Delta p = 3.4\,Pa$. Nevertheless, the numerical solution is still in good agreement with its analytical counterpart. At first glance, the conventional ISPH solution ($\square$) shows a similar saw tooth like error profile. The maximum absolute error is added up to over $4\,Pa$, arising also in the vicinity of the wall. However, the maximum relative error occurs at the free surface ($y/H = 1$), with the pressure at the free surface being unequal to zero. This indicates that the free surface treatment by the simplified virtual particle approach is not sufficient for the conventional ISPH approach. As mentioned in the introduction of this section, a deficient treatment of the free surface results in errors accumulating over time, which can lead to pressure oscillations, as depicted in figure 4.8 after $t = 0.09\,s$ (9,000 time steps). While the absolute error of the NCSPH and NSPH solution remains bounded, the pressure values of particles in the vicinity of the free surface are oscillating with the conventional ISPH approach. In figure 4.9, the evolution of the kinetic energy E_{kin} of the non-viscous fluid column is depicted

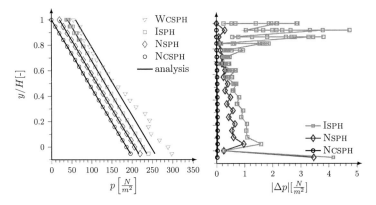

Figure 4.8: Left: Pressure profile after $t = 0.09\,s$ of the NCSPH (o), NSPH (◇), ISPH (□) and WCSPH (▽) approach over the channel height H in comparison with the analytical solution (solid line); Right: absolute error of the respective method with respect to the analytical solution.

over time. At around approximately 10,000 time steps, the kinetic energy starts to rise exponentially, stopping the numerical computation of the ISPH approach. Since, the fluid is treated as non-viscous, no dissipative forces arise during the simulation. One has to stress, that refining the discretization by increasing the number of particles over the channel height does not suppress the numerical instabilities, but increases the numerical accuracy of the solution only. By using the incompressible NSPH approach, the error is effectively reduced, which results in a much more stable numerical scheme. The abort of the simulation due to error accumulation at the free surface can be suppressed for over 600,000 iterations. In case of the incompressible NCSPH approach, no significant error accumulation takes place due to its accurate treatment of free surfaces and solid boundaries. Therewith, the NCSPH scheme is the most stable numerical scheme for systems with negligible dynamics and no prominent oscillations are observed within 1,000,000 time steps as depicted in figure 4.9. Nevertheless, a small error accumulation is observed over time.

In the present section it has been shown, that the conventional incompressible SPH method in conjunction with the simplified virtual particle approach is prone to numerical

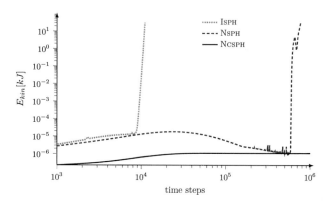

Figure 4.9: Evolution of the kinetic energy in the hydrostatic test case as a function of integration time steps for the NCSPH (–), NSPH (- -) and ISPH approach (· · ·).

instabilities, which are introduced by spurious boundary effects at the free surface. With the adapted incompressible SPH approaches, more accurate and more stable numerical simulations in the presence of free surfaces are possible. By using the kernel renormalization, as done in the NSPH approach and the simplified virtual particle approach, the stability of the numerical scheme is increased. Even better results are obtained using the NCSPH approach, confirming the applicability of the newly introduced Laplacian (equation (3.62)) for discretizing the pressure Poisson matrix. In addition, the gradient of the resulting pressure field is exactly evaluated by the corrected SPH gradient approximation due to its first order consistent nature. Therewith, both methods offer an advantage over the conventional Incompressible SPH approach for the dynamic simulation of materials including interfaces as well as free surfaces at rather low particle dynamics.

Parameter		Value
maximum time step	Δt	$10^{-5}\,s$
initial particle distance	L_0	$1\,ul$
channel height	H	$20\,L_0$
channel width	L	$10\,L_0$
cutoff radius	r_{cut}	$3.1\,L_0$
surface bound	β^S	0.8
initial fluid density	ϱ^0	$1\,\frac{kg}{ul^3}$
gravitational acceleration	g	$9.8\,\frac{ul}{s^2}$
polytropic constant	γ	7
nominal speed of sound	c_0	$1400\,\frac{ul}{s}$

Table 4.2: Computational details of the hydrostatic free surface channel test case. All measures of length are given in chosen units abbreviated by ul as unit length.

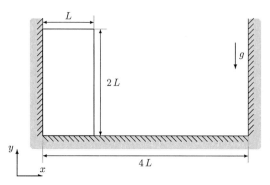

Figure 4.10: Computational domain of the two–dimensional dam break test case.

4.1.3 Collapsing water column after dam break

In the following section, the hydrostatic test case is developed into the dynamic evolution of a slumping water column. As depicted in figure 4.10, a water column is set up on the left side of the basin. The water column is discretized by 648 fluid particles arranged on a 18×36 regular lattice. After removal of the imaginary supporting board, the water column collapses. The generated flow field and the impact of the leading edge of fluid on the right vertical wall has been widely used as a test problem of the applicability of numerical methods for the simulation of large deformations, fluid fragmentation and coalescence [81, 165, 172, 193, 194, 217, 219]. Therefore, the two developed incompressible smoothed particle approaches NSPH and NCSPH are applied to the dam break test case and the results are compared to the one obtained by the state-of-the-art conventional ISPH approach as described in section 2.7.2. In the simulation, the fluid is considered incompressible and inviscid. The computational details are summarized at the end of the section in table 4.3. Furthermore, one has to stress, that arising negative pressure values, which are due to a local decrease in the density field during the simulation (respectively locally negative divergence of the velocity field), are neglected. As soon as the pressure of a fluid particle falls below zero, the pressure is set to zero in the current time step. Without this simplification, the simulation is very unstable and prone to particle clustering, which ultimately stops the simulation. However, Hu and Adams developed a scaling approach, which at least

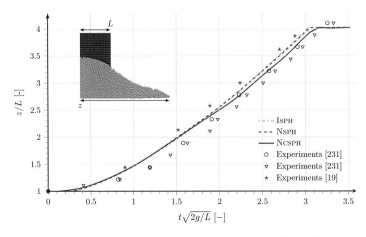

Figure 4.11: Dimensionless front position of the leading edge z/L of a falling water column. Comparison of the front positions simulated with the conventional IsPH, NsPH and NcSPH method with experimental results. The definition of the nondimensional leading edge is given in the upper left detail.

partly alleviates that problem [170].

In figure 4.11 the non-dimensional position of the leading edge of the collapsing water column z/L, as shown in the detail in the figure, is depicted over the non-dimensional simulation time $t^* = t\sqrt{2\,g/L}$ for the conventional IsPH, incompressible NsPH and incompressible NcSPH approach. Experimental results of the collapsing water front of Martin and Moyce ($\nabla, \circ$)[231] as well as Hirt and Nichols ($\star$)[19] are marked for comparison. All three methods are in good agreement with the experimental results, with the NcSPH approach being slightly closer two the experimental results of Martin and Moyce towards the end of the simulation. However, as a trend the leading edge of the water column spreads faster in all three simulation approaches than observed in two of the three experimental results. The deviation can be contributed in parts to the neglect of viscous effects in the simulation and so missing friction of the fluid particles at the wall, even though the collapse of the column is mostly governed by inertia effects. Beside the accurate and quantitative prediction of the process dynamics, a stable and smooth evolution of the free surface represented by the particles is crucial. This is

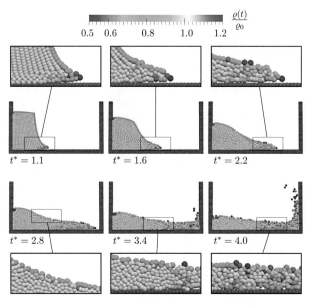

Figure 4.12: Collapse of a water column using the conventional ISPH approach: Particle configuration and respective relative fluid density depicted at the dimensionless time steps $t^* = 1.1, 1.6, 2.2, 2.8, 3.4$ and 4.

especially true for the quantitative simulation of open-porous materials. Consequently, snapshots of the water particle positions at the dimensionless time t^* as indicated in the plots are shown for all discretization methods. The color coding of the particles represents the normalized particle density $\varrho_i(t^*)$ in relation to the initial incompressible density ϱ^0. In figure 4.12, the temporal evolution of the water front from the initial state until the impingement on the vertical wall obtained with the conventional ISPH approach is shown. A particular attention is paid to the particle configuration near the free surface as displayed in the details above the respective plots. Using the conventional ISPH approach, the particle configuration at the free surface stays relatively smooth until the impingement on the boundary, with the largest distortion being observed at the leading edge. The partly irregular configuration also leads to a large decrease of the fluid density at the free surface. At the leading edge, the decrease in density sums

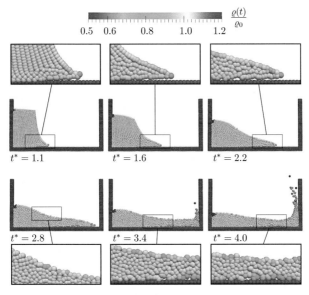

Figure 4.13: Collapse of a water column using the incompressible NSPH approach: Particle configuration and respective relative fluid density depicted at the dimensionless time steps $t^* = 1.1$, 1.6, 2.2, 2.8, 3.4 and 4.

up to 50 % for certain particles. However, the decrease in density at the free surface is not limited to the leading edge, but occurs over the whole free surface with ongoing simulation time.

In figure 4.13 the collapse of the water column simulated with the incompressible NSPH approach is depicted. As expected, the principal dynamics of the collapse is in accordance with the one of the conventional ISPH method. Though, the particle configuration at the free surface is much smoother and more regular in comparison to the conventional approach. In addition, the decrease of the fluid density towards the free surface is by far less pronounced. As long as the particles form a continuum, the departure of the local fluid density from the incompressible density is less than 2 % in the bulk and less than 10 % at the free surface. An exception is the upper left contact point of the free surface with the wall boundary. Due to the simplified treatment of

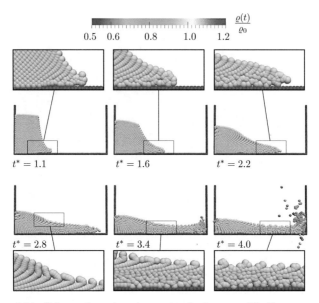

Figure 4.14: Collapse of a water column using the incompressible NCSPH approach: Particle configuration and respective relative fluid density depicted at the dimensionless time steps $t^* = 1.1$, 1.6, 2.2, 2.8, 3.4 and 4.

the wall boundary as described in section 3.2.2.2, the error in the local density exceeds 20 %. This is due to the zero pressure boundary condition at the free surface. At that point, where the solid and the fluid particles are considered as free surface particles, the repulsive forces between these particles are too small, leading to the observed clustering. In figure 4.14, the simulation results of the collapse of the water column using the NCSPH approach are depicted at the previous non-dimensional time steps. And as expected, the fundamental evolution of the water column is in accordance with the simulation results presented previously. However, the dynamically changing particle arrangement at the leading edge and free surface differs from both previous results. In the details of the leading edge shown in the upper row of figure 4.14, the leading edge appears not as spiky as in the results obtained from the conventional and Normalized SPH approaches. Yet, the free surface particle configuration is by far smoother in comparison to the

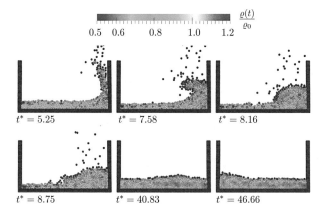

Figure 4.15: Collapse of a water column using the ISPH approach, second period: Particle configuration and respective relative fluid density depicted at the dimensionless time steps $t^* = 5.3$, 7.6, 8.2, 8.8, 40.8 and 46.7

conventional SPH variant, while in comparison to the incompressible NSPH approach, the particles configuration at the free surface appears rougher. This is due to the zero pressure boundary condition at the free surface. In contrast to the former two methods, the zero pressure Dirichlet boundary condition is directly impressed on the respective boundary particles as described in section 3.2.2.1. Therewith, no repulsive force acts between two free surface particles, which leads to the small particle clusters observed in the detail of the next to last and last figure. In the current simulation, the free surface particles are identified using equation (3.86) and (3.87). Nevertheless, the incompressibility of the fluid is ensured with a deviation of the actual fluid density to the initial density by about 2 % for the bulk and surface particles. However, this is only true as long as the particles form a continuum and as long as the particles are not impinging on the opposite vertical wall. As it can be seen in the last figure of the series, the current pressure coupling between solid and fluid particles used in the NCSPH approach is not capable of describing the impingement accurately. Interestingly, all of the fluid particles are decelerated and deflected by the solid wall, but a few fluid particles penetrate the solid wall downwards in vertical direction. This can be explained by the clipping of negative pressure values to zero, which can happen for a negative divergence of the

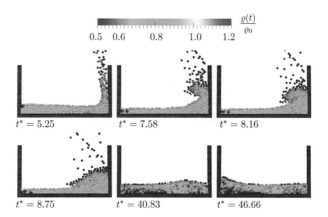

Figure 4.16: Collapse of a water column using the incompressible NSPH approach, second period: Particle configuration and respective relative fluid density depicted at the dimensionless time steps $t^* = 5.3$, 7.6, 8.2, 8.8, 40.8 and 46.7.

particle velocity as described above. And since only one layer of boundary particles is used in the NCSPH approach, the reflective pressure exerted by that layer of boundary particles is not sufficient to prevent the fluid particles from penetrating the solid wall. With ongoing time, more and more particles are dropping through the bottom wall and for this reason the simulation is stopped.

In figure 4.15 and 4.16, the second phase of the collapse after the impingement of the water front on the right vertical wall is shown for the conventional ISPH and incompressible NSPH method. In both cases, the fluid particles swap from one edge to the other until the particles come to rest. Similar to the previously presented results, the particles are more regularly distributed at the free surface for the NSPH approach. This results in a smoother free surface in comparison to the conventional approach. Yet after impingement, a large deviation of the local density from the initial incompressible density is observed for the NSPH approach as shown in figure 4.16. The increasing deviation from the incompressible density is vested in the integration of the volume of each particle. As shown in the conditional equation for the density (equation (3.94)), the volume of the particles is used for the renormalization of the kernel function as presented in section 3.2.2.2. Due to the temporal integration of the divergence of the velocity field, the error in the calculation of the particle volume at each point in time is accumulated over time. As long as the error is small, as in the first phase of the collapse before the impingement on the wall, the normalization of the kernel function results in a higher accuracy for describing the fluid density. But after the impingement, the fluid is fragmented and the error in the divergence of the velocity field increases due to the incomplete integration domain of the sloshing fluid particles. And due to the temporal integration of the volume, the error is retained after the fluid particles converge again in the basin. In contrast, the error in the density field is not summed up in the conventional ISPH approach, since the fluid density is estimated from the actual particle distribution in every time step as shown in equation (2.96).

To conclude, all three methods are able to quantitatively describe the dynamics of the collapsing water column until the impact of the leading edge on the vertical wall. In this first part of the overall process, the flow is characterized by the quick spreading of the liquid and the development of the leading edge. Especially during the spreading of the water front, the free surface particle configuration has to be smooth to guarantee an accurate representation of the free surface. The evolution of the liquid in terms of a smooth free surface particle configuration is most accurately described by the NSPH approach. In contrast, the free surface obtained by the NCSPH approach is relatively smooth in the beginning, while the particle configuration is getting more and more distorted with increasing simulation time. The most irregular particle configuration at the free surface is obtained with the conventional ISPH approach. Regarding the accurate representation of the fluid density during the spreading of the liquid, the incompressibility of the fluid is best conserved by the NCSPH approach. In addition, the fluid density does not decrease towards the free surface. The same observation holds for the results obtained by the NSPH approach, while the incompressibility of the fluid is not as strictly conserved as in the NCSPH method. In contrast, the fluid density calculated with the conventional ISPH approach decreases sharply towards the free surface and the incompressibility of the fluid is not as accurately conserved as in the latter two cases. However, the conventional method is best suited for highly dynamic fluid flows, where fragmentation and recombination of the fluid occurs as illustrated above. Therefore, the advantage of the developed incompressible NSPH as well as NCSPH approach lies in the more accurate and smooth representation of the free surface. The disadvantage of the NSPH approach lies in the accumulation of the error for calculating the fluid density.

Furthermore, the current treatment of the fluid – solid boundary in the Ncsph method is not capable of treating the highly dynamic impingement of the fluid particles on the wall.

In the present section, the ability of the various Sph approaches for describing highly dynamic free surface flows has been demonstrated on a rather qualitative basis. In the following paragraph, the qualification of the developed approaches for simulating free surface flows on a quantitative level will be investigated.

Parameter		Value
maximum time step	Δt	$10^{-4}\,s$
initial particle distance	L_0	$0.08\,ul$
characteristic length	L	$18\,L_0$
cutoff radius	r_{cut}	$2.1\,L_0$
surface bound	β^S	0.98
initial fluid density	ϱ^0	$1\,\frac{kg}{ul^3}$
gravitational acceleration	g	$9.8\,\frac{ul}{s^2}$

Table 4.3: Computational details of the dam break test case. All measures of length are given in chosen units abbreviated by ul as unit length.

4.1.4   Free surface flow of a fluid droplet

In this section, the deformation of an initially circular fluid drop is investigated. This test case was first introduced by Monaghan and is included in the present work to illustrate the ability of the developed incompressible SPH approaches to quantitatively describe free surface flows [49]. Additionally, it will be demonstrated, that the Normalized as well as the Normalized Corrected SPH approach conserves total linear and angular momentum in the absence of external forces. Colagrossi used the test case later to study the effect of artificial viscosity and for convergence studies [67]. Bonet and Lok applied the test case for studying the effect of a renormalized corrected kernel gradient on the conservation properties of SPH [87]. Fang *et al.* studied the impact of the Finite Particle Method, an advancement of the Corrected Smoothed Particle Method introduced in section 2.5, on the accuracy of the droplet evolution [228].

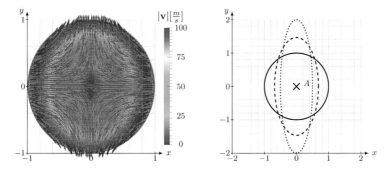

Figure 4.17: Left: Initial particle configuration of the fluid drop and impressed initial velocity field (colored vectors). The color coding represents the absolute value of the velocity vectors. Right: Initial position of the free surface boundary (solid line) and position of the free surface at time instant $t = 0.4/A_0$ (dashed line) and $t = 0.8/A_0$ (dotted line) [67].

Initially, the inviscid fluid drop possesses a circular shape in the two-dimensional domain, with an initial radius R and its center located at the origin ($\boldsymbol{x} = 0, 0$). The following velocity field is impressed on the fluid drop

$$\boldsymbol{v}_0 = \begin{pmatrix} -A_0 x \\ +A_0 y \end{pmatrix}$$

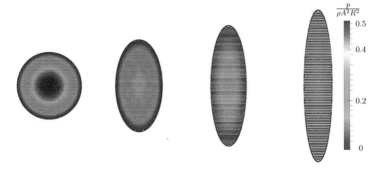

Figure 4.18: Deformation of initially circular fluid drop at the dimensionless points in time $t\,A_0 = 0, 0.03, 0.06, 0.1$. The color coding of the particles display the nondimensional pressure $\frac{p}{\varrho A_0^2 R^2}$.

with x and y being the components of the position vector $\boldsymbol{x}$. The computational domain is shown in figure 4.17 together with the initially impressed velocity field $\boldsymbol{v}_0$. During the course of the calculation, the drop should remain elliptical, with the product of the semi-minor axis a and semi-major axis b remaining constant. Besides, the surface boundary of the drop should remain smooth as stated by Bonet and Lok [87]. For verification of the SPH based approaches, the evolution of the pressure field of the fluid drop is calculated by the analytical expression [67, 228]

$$p^a(\boldsymbol{x}, t) = 0.5\varrho \left[\dot{A}(x^2 - y^2) - A^2(x^2 + y^2) - a^2(\dot{A} - A^2) \right], \qquad (4.4)$$

with A being defined as

$$\frac{d\,A}{d\,t} = \frac{A^2(a^4 - a^2 b^2)}{a^4 + a^2 b^2}. \qquad (4.5)$$

The derivation of the semi-analytical solution is sketched in appendix F and used to verify the proposed incompressible SPH algorithms in the following.

The evolution of the non-viscous, circular fluid drop has been simulated up to time point $t = 1.5/A_0$ using the NSPH and NCSPH approach. In addition, the conventional ISPH solution is used for comparison. The fluid drop, as shown in figure 4.17 is discretized by 5899 particles, distributed in a hexagonal configuration with an initial particle spacing of $L_0 = 0.025\,ul$. The smoothing length is set to $h = 1.55\,L_0$ resulting in a cutoff radius of $r_{cut} = 3.1\,L_0$ for the cubic spline kernel function. All computational

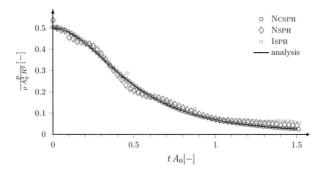

Figure 4.19: Time history of the pressure at the center of the drop. The analytical solution is shown as a black solid line. The numerical solutions are marked with □ in the case of Isph, ◇ for the Nsph and ○ for the Ncsph method.

details are summarized in table 4.6. In figure 4.18, the temporal evolution of the fluid drop, simulated with the Normalized Corrected Sph method, is depicted exemplarily at the time points $t = 0, 0.3/A_0, 0.6/A_0$ and $1/A_0$. The particle color determines the nondimensional pressure $\frac{p}{\varrho A_0^2 R^2}$. As designated, the fluid drop remains in an elliptical shape and the free surface retains smooth over the whole simulation time. In figure 4.19 the pressure in the center of the fluid drop at location A as indicated in figure 4.17 is drawn over the dimensionless time for all three Sph solutions with the semi-analytical solution being included as a reference (solid line). The pressure history obtained with the Isph (□) as well as the Nsph approach (◇) oscillate around the analytical solution with small amplitude. Yet, the Nsph approach shows slightly more accurate results than the conventional Isph solution. In contrast, the incompressible Normalized Corrected Sph approach does not show any pronounced oscillations and yields very accurate results. The maximum deviation of the pressure value of the Ncsph approach from the analytical solution is calculated to $\epsilon = 0.93\,\%$. In contrast, the maximum deviation of the conventional Isph and the incompressible Nsph algorithm reaches several hundred percent due to the pronouced offset towards the end of both simulations. Furthermore, one has to stress, that the pressure oscillations in case of the Isph and Nsph approach are even more pronounced at the free surface as shown in figure 4.20.

To investigate the convergence properties of the developed Sph algorithms, the simulation

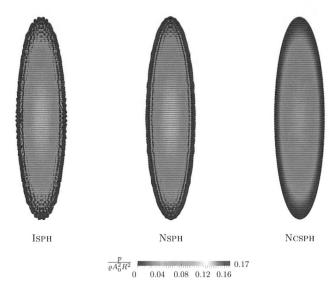

$\frac{p}{\varrho A_0^2 R^2}$ ▦▦▦ 0.17
  0 0.04 0.08 0.12 0.16

Figure 4.20: Deformation of initially circular fluid drop at the dimensionless points in time $t\,A_0 = 0.8$. Left: Conventional ISPH; Center: incompressible NSPH; Right: incompressible NCSPH. The color coding of the particles displays the nondimensional pressure field $\frac{p}{\varrho A_0^2 R^2}$.

is also conducted using $N = 1519$ particles. Therewith, the initial inter-particle distance is extended to $L_0 = 0.05$. The overall error in percent is defined as

$$\epsilon(N, \tau) = \frac{\int_0^\tau |p(N, t) - p^a(t)|\, d\,t}{\int_0^\tau |p^a(t)|\, d\,t} \times 100, \qquad (4.6)$$

with $p(N, t)$ being the pressure estimated at time t using a simulation with N particles. The expression $p^a(t)$ represents the corresponding analytical value. The integration domain is bounded by the point in time $\tau = 1.5/A_0$. As already sketched above, the NCSPH method yields the most accurate results with an overall time-averaged deviation of the pressure value from the analytical solution of 0.27 % using 5899 particles and 0.69 % for the simulation with 1519 particles, respectively. The accuracy of the NSPH approach is similar to the conventional ISPH method, with an overall error around 6 % for NSPH and 8 % for ISPH using the higher and 18 %, respectively 16 % for the lower resolution. The results are summarized in table 4.4, together with the order of convergence o. The latter is defined as

$$o = \frac{\ln\left(\frac{\epsilon(1519,\tau)}{\epsilon(5899,\tau)}\right)}{\ln(2)}, \tag{4.7}$$

with a refinement ratio of two. The conventional Isph approach shows almost linear convergence for the pressure value, while an order of convergence to the analytical solution of greater than one is obtained for the Nsph and Ncsph method.

	$\epsilon(1519,\tau)[\%]$	$\epsilon(5899,\tau)[\%]$	o
Isph	15.7	7.9	0.98
Nsph	18.3	6.2	1.6
Ncsph	0.69	0.27	1.4

Table 4.4: Convergence analysis of the deformation of an initially circular fluid drop. The overall error in percent $\epsilon(N,\tau)$ as defined in equation (4.6) is shown together with the order of convergence o for the conventional Isph, the incompressible Nsph and Ncsph approach.

For describing the evolution of open-porous materials during the manufacturing step, the accurate prediction of deforming surfaces is crucial. Therefore, the deformation of the fluid drop is quantified by the value of the semi-minor axis parameter a and semi-major axis b of the three Sph approaches, which are shown at different time instants in table 4.5 together with the respective analytical value. Despite the prominent pressure oscillations of the Isph and incompressible Nsph method, both approaches describe the evolution of the fluid drop with a very high accuracy. The maximum deviation from the analytical solution adds up to 0.04 % and 0.09 % percent. Perfect agreement is achieved by the Ncsph method, resulting in a maximum deviation of $9.4 \cdot 10^{-4}$ %.

$tA_0[-]$	0.25		0.5		0.75		1.0		1.25	
	a	b	a	b	a	b	a	b	a	b
analytic	0.783	1.278	0.627	1.595	0.518	1.931	0.439	2.276	0.381	2.625
Isph	0.781	1.282	0.621	1.603	0.524	1.938	0.449	2.278	0.380	2.620
Nsph	0.78	1.282	0.626	1.604	0.515	1.944	0.434	2.292	0.373	2.643
Ncsph	0.783	1.278	0.627	1.595	0.518	1.932	0.440	2.278	0.382	2.624

Table 4.5: Comparison of the numerical and theoretical semi-axes a and b over the dimensionless time tA_0.

Even though, the evolution of the semi-axes of the fluid drop are predicted with very high accuracy by all three methods, the shape of the free surface and particle arrangement differs strongly. For good results, the particle configuration has to preserve a smooth outer boundary. In addition, any tendency of fragmentation and breakup should be suppressed. In figure 4.21 the particle configuration of the elliptic drop of the three SPH methods at time instant $t = 0.8/A_0$ is depicted. The color coding of the particles represents the dimensionless fluid density $\frac{\varrho}{\varrho_0}$, with ϱ_0 being the initial density. As one can see, the free surface obtained by the conventional ISPH approach, shown on the left, is strongly ragged over the whole boundary. This is especially pronounced at the upper and lower boundary as shown in the detail of figure 4.21. Besides, the density of the ISPH solution decreases significantly towards the free surface due to the appearing particle deficit. The largest decline of the dimensionless density with $\frac{\varrho}{\varrho_0} = 0.41$ occurs in the strongly raged regions of the free surface. By introducing the kernel renormalization, as done for the NSPH approach, the decline of the density towards the free surface is reduced as shown in center of figure 4.21. Furthermore, the surface of the fluid drop is kept relatively smooth. However, a tendency for fragmentation can be seen as shown in the lower detail of figure 4.21. The quantitatively and qualitatively best results for the deformation of the fluid drop are obtained using the NCSPH based approach as depicted in figure 4.21 on the right. The free surface remains smooth and no significant decline of the density towards the free boundary is observed. But these results have to be settled with increased numerical costs, since the total simulation time for the current example is increased by 37 %.

Last but not least, the conservation properties of the incompressible NSPH and NCSPH approach are investigated and compared to the conventional ISPH method. As long as the discretization schemes for the continuity and momentum balance are derived from a variational principle as shown in section 3.2, the linear and angular momentum should be conserved exactly. But within the NCSPH framework, the variational consistent pressure gradient operator, shown in equation (3.47), is not even zeroth order consistent. Based on that, the non-symmetric, but first order consistent equation (3.77) is used. Hence, the conservation of linear as well as angular momentum of both proposed method is investigated below. In figure 4.22, the time history of the total linear and angular momentum is shown. All three methods preserve the total momenta, which is considered as a good basis for well-behaved and accurate simulations by Bonet and Lok [87].

In the present section, the conservation properties of the two developed numerical schemes have been examined. In addition, a strong focus was laid on the verification of

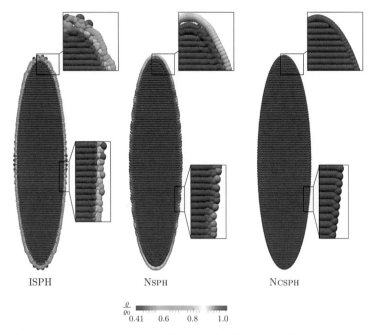

ISPH   Nsph   Ncsph

$\frac{\varrho}{\varrho_0}$

0.41   0.6   0.8   1.0

Figure 4.21: Deformation of the initially circular fluid drop at the dimensionless points in time $t\,A_0 = 0.8$. Left: Conventional Isph; Center: incompressible Nsph; Right: incompressible Ncsph. The color coding of the particles displays the density ratio $\frac{\varrho}{\varrho_0}$.

the quantitative description of the evolution of free surfaces. Even though, accurate results of the free surface evolution under large deformations have been obtained with all three methods, irregularities at the free surface have been observed for the conventional Isph as well as Nsph approach. Since the overall dynamics of the system in the regarded test case is mostly governed by bulk particles, only limited statements about the stability of the free surface itself can be made. Therefore, the dynamics of a system dominated by boundary effects will be studied in the following paragraph.

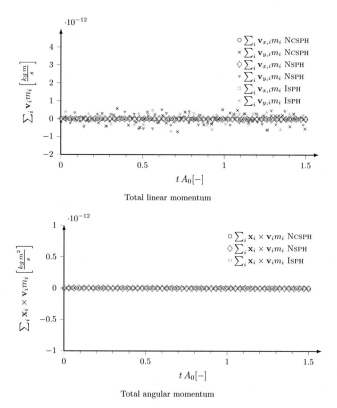

Figure 4.22: Total linear and angular momentum of the deforming elliptic drop displayed over the dimensionless time. The conventional ISPH solution is depicted by ($\square$), the NSPH solution by ($\diamond$) and the NCSPH solution is marked by ($\circ$).

Parameter		Value
initial particle distance large domain	L_0	$0.025\,ul$
initial particle distance small domain	L_0	$0.05\,ul$
cutoff radius	r_{cut}	$3.1\,L_0$
drop radius	R	$40\,L_0$
surface boundary	β^S	0.8
initial fluid density	ϱ^0	$1\,\frac{kg}{ul^3}$
parameter	A^0	$10\,\frac{1}{s}$

Table 4.6: Computational details of the simulation of the evolution of an elliptic non-viscous and incompressible fluid drop. All measures of length are given in chosen units abbreviated by ul as unit length.

4.1.5   Stability of a fluid patch with a free surface

In the following paragraph, the evolution of a rotating, initially square shaped fluid drop is presented. This test case was developed by Colagrossi for assessing artificial tensile stability techniques and reinitialization of the particle density approaches for SPH based numerical schemes [67]. During the course of the simulation, the fluid patch undergoes large free surface deformations, resulting in a negative pressure field. In that way, the test case is challenging with respect to the occurrence of the well-known tensile instability phenomena [75]. Furthermore, the influence of truncated interpolation domains is dominant due to the relatively larger free surface area.

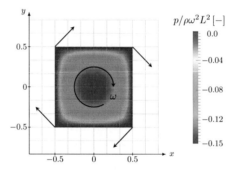

Figure 4.23: Initial particle configuration of the rotating fluid patch. The contours represent the resulting initial pressure field due to the impressed velocity field. In addition the bisecting trajectories of the vertex particles are depicted [67].

At the initial time, the computational domain is composed of a two-dimensional fluid patch with edge length L. The following velocity field is impressed on the fluid patch

$$v_0 = \begin{pmatrix} +\omega y \\ -\omega x \end{pmatrix}, \qquad (4.8)$$

resulting in a pure rotation around its origin with the angular velocity ω as shown in figure 4.23. The fluid patch itself is represented by 41×41 particles, distributed on a regular lattice. In difference to previously discussed test cases, no analytical solution is available. However, some considerations about the dynamic behavior of the fluid patch have been made by Colagrossi and are repeated subsequently [67]. Due to the rotation of the fluid patch and the emerging centrifugal forces, the four corners are elongated,

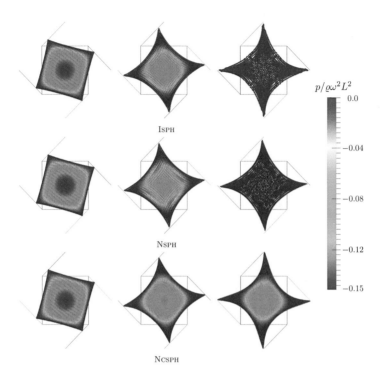

$p/\varrho\omega^2 L^2$

0.0

−0.04

−0.08

−0.12

−0.15

Figure 4.24: Evolution of the initially square fluid drop over time calculated with different incompressible SPH methods. Upper row: Conventional ISPH; Center row: incompressible NSPH approach; Lower row: incompressible NCSPH approach. The particle positions and respective pressure fields are depicted at the dimensionless points in time $t\omega = 0.25, 0.75, 1.0$. The initial size of the fluid drop is sketched as a reference, together with the trajectories of the vertex points.

resulting in a starfish-like structure. In absence of external forces, the pressure gradients of the four vertex particles of the fluid square are zero. These particles remain at the vertex points, preserving their initial velocity and thence, keep moving on the bisecting lines as indicated in figure 4.23. From the beginning of the simulation, the fluid domain is under tension and a negative pressure field develops. As mentioned above, the occurrence of negative pressure values is always problematic for SPH based algorithms, due to the tensile instability phenomena. In addition, the free surface gets extremely deformed during the course of the simulation. Based on that, the test case is challenging for simulation techniques based on the smoothed particle approach and an important test case for the applicability of SPH to describe the formation and evolution of thin structures, e.g. ligaments of a pore matrix. For the simulation of open-porous materials the accurate description of the free surface and the suppression of spurious errors introduced at the boundaries are crucial. For this reason, the applicability of the developed incompressible NSPH and NCSPH methods are investigated and compared to the standard ISPH approach. The computational details for the simulation of the following test case are summarized in table 4.7 at the end of the section.

In figure 4.24, the obtained particle positions of the deforming fluid patch are presented for the three investigated methods at different points in time together with the respective pressure fields. In the upper row (4.24a), the evolution of the fluid patch calculated with the conventional SPH approach is shown, while the particle configuration obtained by the elaborated incompressible NSPH approach are presented in the center row (4.24b). In the lower row (4.24c), the respective particle positions and pressure fields obtained with the incompressible NCSPH method are depicted. At the first point in time ($t\omega = 0.25$), all three approaches yield good results. The particle configuration is regular and smooth and the pressure field shows no oscillations. Besides, the free surface Dirichlet boundary condition of $p = 0$ is (approximately) satisfied by all methods. At the dimensionless time step $t\omega = 0.75$, the first pressure oscillations are observed for the ISPH and NSPH approach. These oscillations seem to originate from spurious boundary effects, which propagate from regions close to the free boundary towards the core region. While the pressure oscillations are more pronounced in case of the NSPH method, the evolution of the fluid corners appears rougher in the conventional ISPH approach. With both methods, the pressure oscillations build up over time, resulting in unphysical high absolute pressure values at the dimensionless time instant $t\omega = 1.0$ as depicted in figure 4.24a and 4.24b. This leads to very large particle velocities, which stop the numerical simulation. Therewith, the NSPH and conventional ISPH approach are not able to describe the evolution of thin structures. In contrast, the pressure field and particle configuration remains stable over the whole simulation time span using the NCSPH method as depicted in figure 4.24c. So, the NCSPH method is the sole investigated method capable of predicting the deformation of the fluid patch and thus of thin structures. The particle configuration at the free surface remains smooth and the movement of the fluid arms is in relatively good agreement with the depicted analytical trajectories as sketched in figure 4.25. However, one has to stress, that the simulation also crashes after the instant in time $t = 2.54/\omega$. Nevertheless, the simulation can be stabilized to a certain extend by increasing the number of particles, which is not the case for the ISPH and NSPH approach.

To conclude, the presented incompressible NSPH approach does not offer any significant improvement in comparison to the conventional ISPH approach, if dealing with the evolution of thin material structures under tension. In both cases, occurring tensile instabilities cause particle clustering, ultimately stopping the numerical computation abruptly. Unlike, the NCSPH approach is the only investigated SPH based approach capable of accurately describing the evolution of thin material structures under these conditions. Therewith, the NCSPH approach is the method of choice for simulating the structure formation of open-porous materials. In addition, one has to stress, that the latter test has been simulated with different (corrected and uncorrected) WCSPH discretization approaches by several authors [67, 226, 228, 232]. In contrast to the present work, an artificial viscosity has been applied in all of the methods published in the cited articles above. Supplemental, Colagrossi and Touze et al. needed to periodically execute a particle remeshing to ensure stability [67, 232].

Parameter		Value
initial particle distance	L_0	$0.025\,ul$
cutoff radius	r_{cut}	$4.1\,L_0$
patch dimension	H	$41\,L_0$
surface boundary	β^S	0.8
initial fluid density	ϱ^0	$1\,\frac{kg}{ul^3}$
angular velocity	ω	$5\,\frac{rad}{s}$

Table 4.7: Computational details of the simulation of the evolution of a non-viscous and incompressible fluid square. All measures of length are given in chosen units abbreviated by ul as unit length.

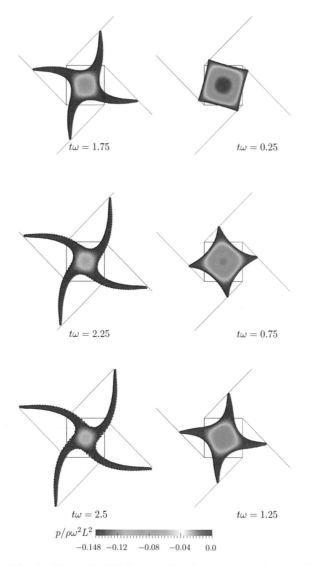

Figure 4.25: Evolution of the initially square fluid drop over time calculated with the incompressible NCSPH method.

4.1.6 Interim conclusion

In the preceding sections, the developed basic incompressible NSPH and NCSPH algorithms for non-viscous fluids have been verified as well as compared to the conventional ISPH method as described in section 2.7.2. It has been shown, that the simplified virtual particle approach, as introduced in section 3.2.2.2, is effective in reducing the errors introduced at boundaries and free surfaces. However, in order to completely restore the particle deficit at free surfaces, the fully corrected and normalized SPH interpolation, as used in the NCSPH approach, is essential. Therewith, the most accurate numerical results as well as most stable numerical scheme is obtained. The latter is especially true if considering the evolution of thin structures as shown in section 4.1.5. In contrast, the NCSPH approach proved to be limited applicable to highly dynamic free surface flows with solid–fluid interactions as presented in section 4.1.3. For these kinds of simulation, the NSPH approach as well as the conventional ISPH approach is better suited. In the subsequent section, the presented numerical schemes are enriched by introducing various material models, which are needed for the simulation of open-porous materials. These models are verified in the following.

4.2 Verification of material models

In the previous sections the focus was laid on assessing the accuracy and stability of the developed SPH algorithms on a general basis, since the accurate representation of free surfaces and thin material structures is a crucial prerequisite for a feasible simulation of the manufacturing process of open-porous materials. Yet, the SPH based approach needs to be also able to accurately describe the rheological behavior of various, partly complex materials for predicting structure evolution during the manufacturing step. Therefore, different material models are investigated in the following, with an emphasis on the validation of the SPH solution in order to guarantee the quantitative solution of the respective material model. With regard to the present application, the suitability of SPH to accurately describe viscoelastic and viscoplastic material behavior is examined. Furthermore, Newtonian fluid behavior is investigated. For describing the deformation of solids, a linear elastic material model is implemented and verified by bending of a cantilever beam under the influence of gravity. However, it is not the scope of this work to find the most realistic material models for the presented application. Quite contrary, a stable and accurate, as well as versatile and consistent numerical approach is developed, in which various material models can be included. As more information about the exact properties of the involved materials is available, more appropriate constitutive equations, which resemble the actual material behavior in a more realistic manner, can be implemented in the present scheme with ease.

In the present paragraph, the computational domain of the so called Poiseuille test case is described. This particular test case has been extensively studied in the literature [22, 79, 103, 104, 106, 121, 233, 234] and it is used in the present work, to show the ability of the presented SPH approaches to describe transient flow of materials with different rheology with high accuracy. Consequently, the numerical results of the two-dimensional transient flow are verified by comparison with the respective analytical solution. Hence, it is important to notice, that all presented simulation results are obtained without using any kind of stabilization techniques, like artificial viscosities or stresses [20, 78]. The computational domain of the Poiseuille channel is shown in figure 4.26. The two stationary plates are positioned at the vertical positions $y = 0$ and $y = H$ and are infinitely extended in the horizontal direction. Initially, the fluid between these plates is at rest and at time point $t > 0$, a body force f is acting on all particles along the channel

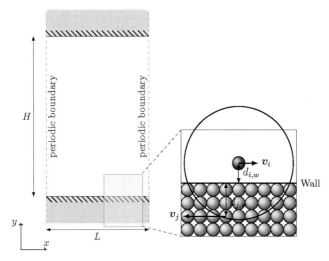

Figure 4.26: Computational domain of the two-dimensional Poiseuille flow test case and detail of the wall near region and mirror imaging to enforce the zero velocity slip at the wall as applied for the conventional SPH as well as renormalized NSPH approach.

axis. For all simulations, the particles are placed on a regular lattice, which leads to a constant initial density for all particles. In horizontal directions, periodic boundary conditions are applied. Thus, SPH particles leaving the boundary are reinserted on the opposite side of the computational domain. As shown in the detail of figure 4.26, the solid walls are represented by fixed particles. For simulations using the conventional ISPH and renormalized NSPH approach, the treatment of the solid walls is based on the work of Morris *et al.* as well as Takeda *et al.* and partly modified for the respective application [22, 23]. Therewith, the no velocity slip boundary condition is enforced on the fixed wall and additional layers of fixed boundary particles as described in section 3.2.2.2. The depth of the reconstructed solid boundary should be at least as large as the chosen cut-off radius of the kernel interpolation. The boundary particles reproduce the desired internal density and are included in the summation to determine any SPH averaged quantity, e.g. velocities or stress gradients. As shown in figure 4.26, for each fluid particle i, the normal distance to the solid boundary $d_{i,w}$ is evaluated. In addition, the distance of the boundary particle $d_{j,w}$ is calculated and its velocity v_j is estimated by extrapolation of the velocity vector of the respective fluid particle $\boldsymbol{v}_i$ over the boundary to

$$\boldsymbol{v}_j = \boldsymbol{v}_w + \frac{d_{j,w}}{d_{i,w}} \left(\boldsymbol{v}_i - \boldsymbol{v}_w \right), \tag{4.9}$$

with $\boldsymbol{v}_w$ being the velocity vector of the solid boundary. In this way, the no-slip boundary condition at the solid boundary is enforced as done by Morris *et al.*[22]. Moreover, if the gradients of the stress tensor have to be calculated explicitly, as it will be shown in the next section, the stress tensor of the boundary particles has to be calculated. This is done by linear extrapolation of the stress tensors of the respective fluid particles across the interface. Therefore, the fluid particle closest to the solid boundary in normal direction, as well as its neighbor particle, which is the second nearest neighbor particle in orthogonal direction to the wall, has to be identified. Based on the stress tensor of these particles and their respective distance to the wall, the values of the stress tensor of the boundary particle are extrapolated. However, this procedure is cumbersome and limited to simple wall geometries. One has to stress, that several other, more versatile and eventually accurate methods exist to enforce the no-slip boundary condition at the solid boundary [22, 23, 81, 106, 109, 132, 135, 155, 233, 235]. Though, the presented pragmatic treatment of the no-slip condition is sufficient within the scope of this work. As mentioned in section 3.2.2.1, the reconstruction of the interpolation domain at the solid boundary is not necessary for the Normalized Corrected SPH approach. In that case it is sufficient to assign the respective velocity, e.g. $\boldsymbol{v}_j = \boldsymbol{v}_w = 0$, to the respective boundary particles.

4.2.1 Viscous material behavior

In the present application, the structure formation is governed by the gaseous blowing agent. Accordingly, the accurate description of viscous fluids is important for modeling the dynamics of the gaseous blowing agent. Since Newtonian fluids play a crucial role in many engineering type fluid dynamics applications, several different approaches for treating fluid viscosity within the SPH framework are proposed in the literature as will be seen later. For Newtonian type fluids, the stress tensor is dependent on the gradient of the velocity field

$$\tau^{\alpha\beta} = \eta \left(\nabla^\alpha \boldsymbol{v}^\beta + \nabla^\beta \boldsymbol{v}^\alpha \right) - \delta^{\alpha\beta} \frac{2}{3} \eta \nabla^\alpha \boldsymbol{v}^\alpha, \tag{4.10}$$

with Stokes assumption of the bulk viscosity $\lambda = -2/3\,\eta$ applied [199]. If the simplification of a spatial constant viscosity coefficient is justified, the divergence of the stress tensor appearing in the momentum balance (equation (3.5)) can be simplified to an expression, which explicitly involves the second order spatial derivatives of the respective velocity field. The divergence of the stress tensor can then be written as

$$\nabla \cdot \boldsymbol{\tau} = \eta \begin{pmatrix} \frac{4}{3}\frac{\partial^2 v_x}{\partial x^2} + \frac{\partial^2 v_x}{\partial y^2} + \frac{\partial^2 v_x}{\partial z^2} + \frac{1}{3}\frac{\partial^2 v_y}{\partial x \partial y} + \frac{1}{3}\frac{\partial^2 v_z}{\partial x \partial z} \\ \frac{4}{3}\frac{\partial^2 v_y}{\partial y^2} + \frac{\partial^2 v_y}{\partial x^2} + \frac{\partial^2 v_y}{\partial z^2} + \frac{1}{3}\frac{\partial^2 v_x}{\partial x \partial y} + \frac{1}{3}\frac{\partial^2 v_z}{\partial y \partial z} \\ \frac{4}{3}\frac{\partial^2 v_z}{\partial z^2} + \frac{\partial^2 v_z}{\partial x^2} + \frac{\partial^2 v_z}{\partial y^2} + \frac{1}{3}\frac{\partial^2 v_x}{\partial x \partial z} + \frac{1}{3}\frac{\partial^2 v_y}{\partial y \partial x} \end{pmatrix}. \tag{4.11}$$

As mentioned in section 2.2.2, the use of the so called hybrid approach, a combination of the SPH first order spatial derivative and a gridless finite difference scheme, is favored within the SPH framework for approximating the second order spatial derivative due to its accuracy, stability and computational efficiency [22, 79, 113, 116, 117, 120, 121, 236–238]. Within the desired application, the reaction-induced structure formation by release of an blowing agent, the gaseous blowing agent is treated as a Newtonian fluid with a spatial constant viscosity. So, the hybrid discretization schemes for the second order derivative for the NSPH (equation 3.101) and NCSPH (equation 4.14) are applied. However, structure formation is governed by the deformation of materials with complex rheology. For this reason, the assumption of a fluid with constant viscosity is not justified in most cases and formulations for e.g. shear-dependent viscosities are validated in addition. This can be seen as a look ahead for modeling viscoelastic and viscoplastic material behavior as presented in section 4.2.2 and 4.2.3. The divergence of the deviatoric stress tensor $\nabla \cdot \tau$ is evaluated as presented in section 3.2, using equation (3.99) for the NSPH and equation (3.54) for the NCSPH approach. The gradient of the velocity vector in the expression for the stress tensor is discretized using equation (3.91) for the NSPH and equation (3.44) for the NCSPH approach. For the conventional SPH solution, which is used as a reference, spatial gradients are discretized by equation (2.39). While the nested gradient approach is more versatile, it is known in the literature, that the evaluation of the second order derivative by two nested gradient operators can lead to oscillations due to the larger extent of the computational stencil [100, 120]. Nevertheless, the nested summation approach is used for modeling heat conduction [95] and viscous flow by many authors [83, 87, 93, 94].

For validation purposes, the regarded flow in between the stationary wall is considered as incompressible, which further simplifies the expression of the deviatoric stress tensor to

$$\tau^{\alpha\beta} = \eta \left(\nabla^\alpha \boldsymbol{v}^\beta + \nabla^\beta \boldsymbol{v}^\alpha \right). \tag{4.12}$$

Based on that, the results obtained from the three SPH variants can be compared with an analytic solution. The time-dependent solution in series of the velocity component parallel to the wall $v_x(t)$ is written as [22]

$$v_x^a(t) = \frac{\boldsymbol{f}_x \varrho}{2\eta} y(y - H) + \sum_{n=0}^{\infty} \frac{4fH^2\varrho}{(2n+1)^3 \pi^3 \eta} \sin\left((2n+1)\pi\frac{y}{H}\right) e^{-(2n+1)^2 \pi^2 \frac{\eta}{\varrho H^2}t}. \tag{4.13}$$

The latter equation characterizes the transient behavior of a fluid initially at rest towards the well-known steady state parabolic velocity profile. Therein, the dimensionless channel is represented by y, which is defined on the interval $y = [0, H]$ and the body force acting on the fluid along the channel length is abbreviated by $\boldsymbol{f}_x$. The component tangential to the wall of the velocity vector $v_y(t)$ is zero throughout the calculation.

As shown in figure 4.26, the height of the channel H is set to unity, while the horizontal extent L can be arbitrarily chosen due to the periodic boundary conditions in the direction of flow. The value of the dynamic viscosity and the initial density of the particles is given in table 4.11. The cutoff radius for the kernel interpolation is set to $r_{cut} = 2.1\,L_0$ for the NCSPH and $r_{cut} = 2.5\,L_0$ for the ISPH and NSPH based simulation, with L_0 being the initial particle spacing. With the chosen cutoff-radius, approximately 13 respectively 20 neighbor particles are involved in the interpolation. The fluid domain is discretized by 40 particles over the channel height and 11 fluid particles along the direction of flow. Even though the velocity profile does not change along the horizontal axis, at least a minimum number of three rows of particles in the x-direction have to be used to exclude artificial effects due to self-interaction over the periodic boundary condition. In the present work, that number is increased to 11 particles in horizontal direction, resulting in a computational domain of 440 fluid particles in total. For the ISPH and incompressible NSPH based simulations 66 boundary particles are used, while 22 boundary particles are sufficient using the Normalized Corrected NCSPH variant. The particles are distributed on a regular lattice for all simulations. The maximum time step is restricted to $\Delta t = 10^{-4}\,s$. Initially, the fluid particles are at rest and a constant body force $\boldsymbol{f}_i$ is acting on all fluid particles. The body force $\boldsymbol{f}_i$ is chosen in such a way to achieve the desired fluid regime characterized by the Reynolds number *Re*. All computational details are summarized in table 4.11 below.

In figure 4.27, the velocity profiles obtained with the NSPH and NCSPH based approach are depicted over the dimensionless channel height for a relatively low Reynolds number of $Re = 1.25 \cdot 10^{-3}$. For the divergence of the deviatoric stress tensor $\boldsymbol{\tau}$, the nested gradient approach as stated above is used. The particle velocities obtained by the NSPH based simulation are shown in the lower half of the channel, indicated by a red triangle (∇), while the velocity profile obtained using the NCSPH variant are marked by a blue circle ($\bigcirc$) in the upper half. The analytical reference solution is depicted by the black solid line. The velocity profiles shown in figure 4.27 are extracted at the time instants $t = 0.025\,s, 0.05\,s, 0.1\,s, 0.2\,s$ as well as $1.0\,s$ and are in very good agreement with the analytical solution. The same simulation is repeated using the hybrid discretization scheme for the second order derivative, written as

$$(\nabla \cdot \eta \nabla)\boldsymbol{v} = \sum_j \frac{2\eta_i \eta_j}{\eta_i + \eta_j} \frac{2V_j(\boldsymbol{x}_j - \boldsymbol{x}_i) \cdot \widetilde{\nabla}_i \widehat{w}_{ij}}{r_{ij}^2}(\boldsymbol{v}_j - \boldsymbol{v}_i) \qquad (4.14)$$

for the NCSPH method. In case of the NSPH approach, the hybrid scheme reads as follows

$$(\nabla \cdot \eta \nabla)\boldsymbol{v} = \sum_j \frac{2\eta_i \eta_j}{\eta_i + \eta_j} V_j \left(\frac{1}{\Gamma_j} + \frac{1}{\Gamma_i}\right) \frac{(\boldsymbol{x}_j - \boldsymbol{x}_i) \cdot \nabla_i w_{ij}}{r_{ij}^2}(\boldsymbol{v}_j - \boldsymbol{v}_i). \qquad (4.15)$$

Again, very good agreement between both SPH based solutions and the analytical solution can be observed in figure 4.28. This indicates that all four versions are capable to handle transient fluid behavior with very high accuracy.

For a more detailed comparison of the two discretization approaches and SPH variants, the root mean square error is calculated. In order to study the impact of the discretization technique as well as boundary treatment approach, two differently normalized root mean square errors are calculated in the following. First, the local normalized root mean square error (LNRMSE), which is normalized by the local analytical velocity, is defined as

$$\epsilon(t) = \sqrt{\frac{1}{N}\sum_i^N \left(\frac{\boldsymbol{v}_{x,i}(t) - v_x^a(y_i,t)}{v_x^a(y_i,t)}\right)^2} \times 100, \qquad (4.16)$$

with $v_x^a(y_i,t)$ being the velocity obtained from the analytic equation (4.13) at the respective particle position y_i and N the number of fluid particles. The LNRMSE

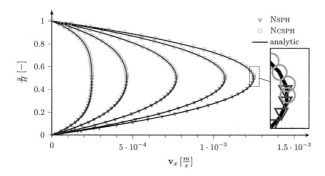

Figure 4.27: Poiseuille flow of a Newtonian fluid discretized by two nested gradients: Comparisons of the velocity profiles between the analytical solution and the numerical NSPH and NCSPH results for $Re = 1.25 \cdot 10^{-3}$. The velocity profiles are shown at the time instants $t = 0.025, 0.05, 0.1, 0.2$ and $1.0\,s$.

for the conventional ISPH, the Normalized SPH and the Normalized Corrected SPH approaches are concentrated in table 4.8. The nested gradient discretization version is abbreviated by $\nabla \cdot \nabla$ and the hybrid Laplacian version by ∇^2 within the table. As

$\epsilon(t)[\%]$	ISPH		NSPH		NCSPH	
$t[s]$	$\nabla \cdot \nabla$	∇^2	$\nabla \cdot \nabla$	∇^2	$\nabla \cdot \nabla$	∇^2
0.025	4.9	2.6	5.0	2.6	2.3	0.8
0.05	3.5	1.8	3.5	1.8	1.6	0.6
0.1	2.4	1.3	2.4	1.3	1.2	0.5
0.2	1.7	1.0	1.7	1.0	1.0	0.4
1.0	1.0	0.9	1.0	0.9	0.9	0.4

Table 4.8: Normalized root mean square error of the horizontal velocity in the channel with a Reynolds number of $Re = 1.25 \cdot 10^{-3}$ at different time points. The LNRMSE is calculated after equation (4.16) and set in relation to the analytical solution shown in equation (4.13).

expected, the LNRMSEs are relatively high due to the zero-crossing of the velocity at the boundary. This is especially true at the startup of the flow ($t = 0.025\,s$), with a LNRMSE of about 5 % for the nested gradient version of the conventional ISPH and incompressible NSPH approach. The NCSPH approximation performs considerably better

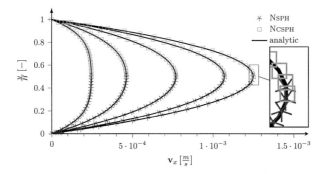

Figure 4.28: Poiseuille flow of a Newtonian fluid discretized by the hybrid second order derivatives: Comparisons of the velocity profiles between the analytical solution and the numerical NSPH and NCSPH results based on the Laplacian for $Re = 1.25 \cdot 10^{-3}$. The velocity profiles are shown at the time instants $t = 0.025, 0.05, 0.1, 0.2$ and $1.0\,s$.

with an error of 2.3 %. With increasing time and absolute velocities, the error is reduced resulting in values around $\epsilon = 1\,\%$ in the steady state. The observation also holds for the Laplacian based discretization, with an overall smaller LNRMSE. Therewith, it can be seen, that the no-slip boundary condition is best enforced within the NCSPH approach as compared to the mirror-imaging approach applied in the conventional SPH as well as Normalized SPH approach. As it will be seen later, the accurate treatment of the no-slip velocity boundary condition at the fluid-solid interface is crucial to maintain a stable simulation scheme for more complex fluid behavior.

For the purpose of assessing the overall actuary of the investigated SPH schemes across the channel height, the root mean square error is slightly modified by normalizing the error with range of the observed velocity $\max(v_x^a(t)) - \min(v_x^a(t))$. This results in the following expression for globally normalized root mean square error (GNRME):

$$E(t) = \frac{\sqrt{\frac{\sum_i^N (v_{x,i}(t) - v_x^a(y_i,t))^2}{N}}}{\max(v_x^a(t)) - \min(v_x^a(t))} \times 100 \tag{4.17}$$

The globally normalized root mean square errors $E(t)$ of the three different simulation methods and two different discretization methods are shown in table 4.9. As it can be seen, the errors $E(t)$ decreases to less than 1 % for all methods. To conclude, all three methods are capable of describing low Reynolds number flow with very high accuracy.

$E(t)[\%]$	Isph		Nsph		Ncsph	
$t[s]$	$\nabla \cdot \nabla$	∇^2	$\nabla \cdot \nabla$	∇^2	$\nabla \cdot \nabla$	∇^2
0.025	0.5	0.3	0.5	0.3	0.5	0.2
0.05	0.3	0.2	0.3	0.2	0.3	0.1
0.1	0.2	0.1	0.2	0.2	0.2	0.1
0.2	0.1	0.1	0.1	0.1	0.2	0.1
1.0	0.1	0.1	0.1	0.1	0.3	0.1

Table 4.9: Global normalized root mean square error of the horizontal velocity in the channel with a Reynolds number of $Re = 1.25 \cdot 10^{-3}$ at different time points. The NRMSE is calculated after equation (4.17) and set in relation to the analytical solution shown in equation (4.13).

In general, the hybrid Laplacian approach yields slightly more accurate results, while no large difference between the different Sph approaches can be observed on a global view. Yet, it has been demonstrated, that the simple treatment of the no-slip velocity boundary condition within the Ncsph approach yields more accurate results in comparison to the more complex mirror-imagine techniques needed for the conventional Sph as well as Nsph approach.

In the following paragraph, the applicability of the presented Nsph and Ncsph approaches to flows with a moderate Reynolds number of $Re = 1.25$ is investigated. Therefore, the simulations described above are repeated with an increased body force acting on all particles. Furthermore, the maximum allowable time step is reduced to $\Delta t = 10^{-7}\,s$. The respective global normalized root mean square errors of all six simulations are presented in table 4.10. All simulations describing the divergence of the stress tensor with the Laplacian hybrid derivative yield very accurate results. The same observation holds for the nested gradient approximation of the second order derivative.

To conclude, all investigated six combinations of the three Sph variants with the two discretization schemes for the divergence of the stress tensor yield very accurate results for the simulation of low to moderate Reynolds number flow. Only minor differences are observed between the different discretization approaches in the current test case. Besides yielding the most accurate results, the Ncsph approach has the remarkable advantage of directly enforcing the respective boundary conditions on the interface. Thus, the reconstruction of the interpolation domain is obsolete. This is of special importance when the deformation of fluids with complex material behavior is investigated, since

$E(t)[\%]$	ISPH		NSPH		NCSPH	
$t[s]$	$\nabla \cdot \nabla$	∇^2	$\nabla \cdot \nabla$	∇^2	$\nabla \cdot \nabla$	∇^2
0.025	0.5	0.3	0.5	0.3	0.5	0.2
0.05	0.3	0.2	0.3	0.2	0.3	0.1
0.1	0.2	0.1	0.2	0.2	0.2	0.1
0.2	0.1	0.1	0.1	0.1	0.2	0.1
1.0	0.1	0.1	0.1	0.1	0.2	0.1

Table 4.10: Global normalized root mean square error of the horizontal velocity in the channel with a Reynolds number of $Re = 1.25$ at different time points. The NRMSE is calculated after equation (4.17) and set in relation to the analytical solution shown in equation (4.13).

for these materials the stress tensor is not only a function of the actual shear rate, but also a function of time as will be shown in section 4.2.3 for the flow of viscoelastic fluids. In the following section, the presented numerical schemes based on the nested gradient approximation for the divergence of the stress tensor are used to describe the deformation of viscoplastic materials by introducing a shear-dependent viscosity.

Parameter		Value
initial particle distance	L_0	$0.025\,ul$
maximum time step $(Re = 1.25 \cdot 10^{-3})$	Δt	$10^{-4}\,s$
maximum time step $(Re = 1.25)$	Δt	$10^{-7}\,s$
cutoff radius NCSPH	r_{cut}	$2.1\,L_0$
cutoff radius ISPH \ NSPH	r_{cut}	$2.5\,L_0$
channel height	H	$40\,L_0$
initial fluid density	ϱ^0	$1\,\frac{kg}{ul^3}$
dynamic viscosity	η	$1\,\frac{kg}{ul\,s}$

Table 4.11: Computational details of the simulation of the Poiseuille flow of an incompressible Newtonian fluid. All measures of length are given in chosen units abbreviated by ul as unit length.

4.2.2 Viscoplastic material behavior

As already stressed above, the generation of open-porous materials is governed by large deformation of complex materials. Especially for the regarded application, the generation of a secondary transport pore system in a polymeric material, viscoplastic material behavior has to be described. According to Hammer, the backbone polymer of the adsorbent monolith behaves like a viscoplastic Bingham material [9] and the Bingham model is known to be capable of describing the rheological behavior of many shear-thinning materials at low shear rates [239]. Based on that, the viscoplastic material behavior is introduced via a constitutive equation in the ordinary momentum balance of the Normalized SPH and Normalized Corrected SPH methods discussed in section 3.1. In the presented work, a regularized Bingham model developed by Papanastasiou*et al.* [240] is chosen due to its numerical stability. Hence, the implementation of a regularized Bingham model within the presented SPH framework is verified by comparison with an analytic solution. One has to stress, that the same model has been implemented using the conventional SPH approach by Zhu *et al.* for studying flow in cylindrical rheometers [234] as well as Martys *et al.* for modeling the flow of a non-Newtonian suspension. So far, the emphasis of the present investigation lies in the development of a stable and accurate numerical algorithm for studying viscoplastic material behavior. If conceivable in the future, more realistic viscoplastic constitutive models can be implemented in the developed framework with ease. For example, a considerably more complex elasto-thermo-viscoplastic material behavior has been implemented within the Modified SPH scheme by Batra and Zhang [129].

By definition, Bingham plastics are characterized by a finite yield stress. As long as the shear rate undercuts the yield stress, the fluid viscosity is infinite, inhibiting viscous flow [241]. As soon as the yield stress is exceeded by the local shear strain rate, Bingham plastics begin to flow. Due to the shear-thinning behavior, the stress tensor depends on the shear strain rate in a non-linear way. Therewith, the divergence of the stress tensor has to be discretized using the nested gradient scheme investigated in detail in the section above. For an ideal Bingham fluid, the constitutive relation is expressed as [242]

$$
\begin{aligned}
\boldsymbol{\tau} &= \left(\eta + \tfrac{\tau_0}{\dot{\gamma}}\right)\dot{\boldsymbol{\epsilon}} \quad \text{for } |\boldsymbol{\tau}| > \tau_0 \\
\dot{\boldsymbol{\epsilon}} &= 0 \quad \text{for } |\boldsymbol{\tau}| < \tau_0,
\end{aligned}
\tag{4.18}
$$

with $\dot{\gamma}$ being the shear strain rate, η the plastic viscosity and τ_0 the yield stress. The shear rate $\dot{\gamma}$ is defined by the second invariant of the deformation strain tensor, which is given by

$$\dot{\epsilon}^{\alpha\beta} = \nabla^\beta v^\alpha + \nabla^\alpha v^\beta. \tag{4.19}$$

The local magnitude of the shear rate is then calculated to

$$\dot{\gamma} = \sqrt{\frac{1}{2}(\dot{\epsilon} : \dot{\epsilon}^T)} = \sqrt{\frac{1}{2}\sum_\alpha \sum_\beta \dot{\epsilon}^{\alpha\beta} \dot{\epsilon}^{\alpha\beta}}. \tag{4.20}$$

Due to the intrinsic discontinuity in the constitutive relation of Bingham plastics, the original Bingham model is difficult to apply in numerical simulation directly [217]. This is especially of concern right after exceeding the viscoplastic yield stress, since the shear rate might be very small at that time point. And since the shear rate also emerges in the denominator of the material model, the apparent viscosity is likely to diverge [234]. Based on these considerations, Papanastasiou suggested a regularized Bingham model, in which these numerical difficulties are circumvented, since the stress tensor $\boldsymbol{\tau}$ is obtained by the continuous equation [240]

$$\boldsymbol{\tau} = \left(\eta + \tau_o \frac{1 - e^{-m\dot{\gamma}}}{\dot{\gamma}}\right)\dot{\epsilon}, \tag{4.21}$$

with m being the parameter marking the transition region between the solid and fluid regimes. A higher value of m results in a sharper transition and vice versa. The choice of the parameter m is dependent on the dimensionless growth parameter M, which is defined as

$$M = \frac{m\, v_{ref}}{H}. \tag{4.22}$$

The occurring parameter v_{ref} is called reference velocity and is set to the value of the maximum velocity occurring for the flow of a purely Newtonian fluid in the respective geometry. According to Chatzimina et al. as well as Zhu et al. the parameter m should be chosen in such a way to obtain a dimensionless growth parameter M in the range of $M \approx 500$ [234, 243]. Due to the sharping of the transition region, a larger growth parameter M leads to smaller time steps after reaching a critical Bingham number [243]. The dimensionless Bingham number is defined as the ratio of yield stress to viscous stress

$$Bn = \frac{\tau_0\, H}{\eta\, v_{ref}}, \tag{4.23}$$

171

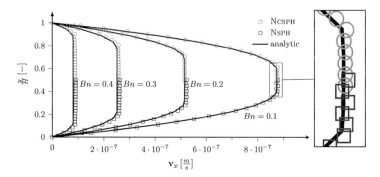

Figure 4.29: Poiseuille flow of different Bingham fluids. Steady state velocity profiles of Bingham fluids characterized by $Bn = 0.4$, $Bn = 0.3$, $Bn = 0.2$ and $Bn = 0.1$ in comparison to the analytical solution. The velocity profile obtained by the NCSPH method are marked with $\bigcirc$, while the NSPH results are highlighted by $\square$. Magnification: Detailed view of the oscillatory behavior of the particle velocities.

and used for the characterization of the flow regimes.

In the following section, the SPH solutions of the Poiseuille flow test case for the regularized Bingham model are verified with the respective analytical solution. The steady-state analytical solution of the Poiseuille flow of a (non-regularized) Bingham fluid is given by Zhu *et al.* as [234]

$$v_x^a = \begin{cases} \frac{1}{2}b\left(\frac{1}{4} - y^2\right) - Bn\left(\frac{1}{2} - y\right) & \text{for } 0.5 \geq y \geq Bn/b \\ \frac{1}{2}b\left(\frac{1}{2} - \frac{Bn}{b}\right)^2 & \text{for } 0 \leq y < Bn/b \end{cases}, \quad (4.24)$$

where b is the dimensionless pressure gradient and y the dimensionless channel height counted from the centerline ($0 \leq y \leq 0.5$).

In figure 4.29, the numerical steady-state velocity profiles over the channel height are depicted and compared to the respective analytical solution. The velocity profiles are computed using the NSPH and NCSPH approach for four Bingham fluids, which are characterized by Bingham numbers of $Bn = 0.1, 0.2, 0.3$ and 0.4. In the upper half of the channel, the NCSPH solution, marked by $\bigcirc$, is shown while the NSPH solution is sketched in the lower part by the red squares $\square$. The cutoff radius for the kernel interpolation is kept constant at $r_{cut} = 2.1 \, L_0$ for the NCSPH approach. In comparison to the purely Newtonian flow presented above, the cutoff radius of the NSPH method is increased from $r_{cut} = 2.5\,L_0$ to $r_{cut} = 3.1\,L_0$ to ensure convergence. The maximum allowable time step size is set to $\Delta t = 10^{-6}\,s$. Further computational details are summarized at the end of the section in table 4.13. As one can see in figure 4.29, the results for the NCSPH and NSPH approach are in good agreement with the analytical solution for all Bingham numbers, while more accurate results are obtained by the Normalized Corrected SPH approach. It is also notable, that more accurate results are obtained for smaller Bingham numbers as it can be seen by the calculated normalized root mean square errors summarized in table 4.12. The normalized root mean square error increases significantly with Bingham numbers larger than $Bn = 0.3$, which can be explained by the increasingly steeper velocity gradient at the transition region. In addition, small oscillations in the velocity profile emerge for all Bingham numbers in the elastic regime of the flow field as shown in the detail in figure 4.29. These oscillations become larger with increasing Bingham numbers and are especially pronounced for the NSPH approach. These oscillations are the reason for increasing the cutoff radius from $2.5\,L_0$ for the NSPH approach in the case of purely Newtonian flow to $3.1\,L_0$ for the viscoplastic flow as stated above. This results in a smoothing length to particle size ratio of $h/L_0 = 1.55$ if using a cubic spline kernel (equation (2.34)). As expected from the results of the previous section, the conventional ISPH method possesses a similar accuracy as the NSPH approach and is therefore not shown.

Bn	$E(t)$ NSPH [%]	$E(t)$ NCSPH [%]
0.1	0.2	0.3
0.2	1.0	0.4
0.3	3.3	2.2
0.4	10.1	5.9

Table 4.12: Normalized root mean square error of the steady state velocity over the channel for fluids with different Bingham numbers Bn. The NRMSE is calculated after equation (4.17) and set in relation to the analytical solution shown in equation (4.24).

To conclude, both developed incompressible SPH approaches are capable of describing transient viscoplastic material behavior with high accuracy. However, the NCSPH approach does not show a pronounced oscillatory behavior and yields more accurate results for all investigated Bingham numbers. Yet, a ratio of smoothing length to initial particle spacing of $h/L_0 = 1.05$ using the cubic spline kernel is sufficient for the NCSPH approach, while the ratio has to be increased to $h/L_0 = 1.55$ to suppress prominent oscillations and achieve accurate results for the NSPH approach. Nevertheless, this is still an acceptable value in terms of computational efficiency, even for three-dimensional applications. Besides, the computation of the corrective terms in the NCSPH method does also not come for free in terms of computational costs. In the next section, the complexity of the rheological model will be further increased by including time-dependent stress–strain effects, therewith enabling the transient simulation of viscoelastic materials.

Parameter		Value
initial particle distance	L_0	$0.025\,ul$
maximum time step	Δt	$10^{-6}\,s$
cutoff radius NCSPH	r_{cut}	$2.1\,L_0$
cutoff radius ISPH \ NSPH	r_{cut}	$3.1\,L_0$
channel height	H	$40\,L_0$
initial fluid density	ϱ^0	$1\,\frac{kg}{ul^3}$
dynamic viscosity	η	$1\,\frac{kg}{ul\,s}$
parameter	m	$5.2 \cdot 10^7$

Table 4.13: Computational details of the simulation of the Poiseuille flow of an incompressible Bingham fluid. All measures of length are given in chosen units abbreviated by ul as unit length.

4.2.3 Viscoelastic material behavior

In the following section, the presented numerical schemes are expanded to the simulation of viscoelastic materials. Many polymeric materials exhibit a stress–strain relation which is not only non-linear, as shown above, but also dependent on the past strain history. Similar to the inclusion of the previous material models, the viscoelasticity is implemented as a constitutive equation in the momentum balance. However, due to its elastic nature, a part of the deviatoric stress tensor $\boldsymbol{\tau}$ has a differential form. Before going into the computational details, some remarks on the choice of the constitutive equation are made, since several viscoelastic material models are available in the literature [199]. The Oldroyd-B fluid constitutive equation has been previously investigated within the conventional SPH context by Ellero *et al.* [106]. In the present work, the Oldroyd-B fluid constitutive equation is implemented within the NSPH and NCSPH framework due to the following reasons. First, the Oldroyd-B fluid model can be reduced to the so called Upper Convected Maxwell (UCM) model, which shows a fully elastic material behavior, with a complex non-linear behavior. Consequently, the UCM model is severely stressing the numerical approach. Second, the Oldroyd-B fluid model is capable of describing polymer flow at least on a qualitative basis. Moreover, it can be augmented to the Giesekus model, which is capable of describing polymer flow on a quantitative basis by proper parameterization [199, 244, 245].

Constitutive equation

As shown before, the overall stress tensor $\boldsymbol{\sigma}$ can be split up in the ordinary isotropic pressure p and the deviatoric or extra stress tensor $\boldsymbol{\tau}$ as shown below

$$\sigma^{\alpha\beta} = -p\,\delta^{\alpha\beta} + \tau^{\alpha\beta}. \tag{4.25}$$

The Oldroyd-B fluid rheological model is used to close the system of equations, with the constitutive equation for $\boldsymbol{\tau}$ given by

$$\tau^{\alpha\beta} + \lambda_1 \overset{\nabla}{\tau}{}^{\alpha\beta} = \eta \left(\dot{\epsilon}^{\alpha\beta} + \lambda_2 \overset{\nabla}{\dot{\epsilon}}{}^{\alpha\beta} \right). \tag{4.26}$$

The parameter λ_1 is called the time constant of relaxation, λ_2 the time constant of retardation and η the total fluid viscosity. The symbol ∇ denotes the upper-convected derivative and $\overset{\nabla}{\tau}{}^{\alpha\beta}$ is defined by [199]

$$\overset{\nabla}{\tau}{}^{\alpha\beta} = \frac{D\,\tau^{\alpha\beta}}{D\,t} - \nabla^\gamma v^\alpha \tau^{\gamma\beta} - \nabla^\gamma v^\beta \tau^{\gamma\alpha}, \tag{4.27}$$

and the symmetric strain tensor $\dot{\epsilon}^{\alpha\beta}$ is defined as

$$\dot{\epsilon}^{\alpha\beta} = \left(\nabla^\alpha v^\beta + \nabla^\beta v^\alpha \right). \tag{4.28}$$

It is common in the literature to split the extra stress tensor $\boldsymbol{\tau}$ of the viscoelastic Oldroyd-B fluid rheological model in

$$\tau^{\alpha\beta} = \eta \frac{\lambda_2}{\lambda_1} \dot{\epsilon}^{\alpha\beta} + \tau_e^{\alpha\beta}, \tag{4.29}$$

where the first expression represents the viscous part and $\tau_e^{\alpha\beta}$ the additional elastic stress tensor [246, 247]. The latter is governed by the following differential equation

$$\tau_e^{\alpha\beta} + \lambda_1 \overset{\nabla}{\tau_e}{}^{\alpha\beta} = \eta \left(1 - \frac{\lambda_2}{\lambda_1} \right) \dot{\epsilon}^{\alpha\beta}. \tag{4.30}$$

By introducing the retardation ratio β as

$$\beta = \frac{\lambda_1 - \lambda_2}{\lambda_1} \tag{4.31}$$

the two limiting cases of the Oldroyd-B fluid model can be obtained [208]. By choosing λ_1 and λ_2 to give $\beta = 0$, equation (4.29) and (4.30) describes a Newtonian fluid. For $\beta = 1$, the purely elastic Upper-Convected Maxwell model is obtained.

In this section, simulation results of the two-dimensional channel flow of a viscoelastic material are presented. The computational domain of the Poiseuille test case is shown at the beginning of section 4.2 in figure 4.26. In order to elaborate the accuracy of the presented SPH approaches, the analytical solution for the startup of a Poiseuille flow in a 2D channel needs to be calculated. Therefore, the dimensionless Weissenberg number Wi is introduced as the product of the characteristic relaxation time of the fluid and the characteristic rate of deformation. In case of the Poiseuille flow, the Weissenberg number is given by

$$Wi = \frac{2v_0\lambda_1}{H}, \tag{4.32}$$

with v_0 being the centerline velocity in the channel. Furthermore, the elasticity number El is introduced as the ratio of the Weissenberg number Wi to the Reynolds number Re to give

$$El = \frac{2\eta\lambda_1}{\varrho H^2}, \tag{4.33}$$

which characterizes the ratio of elastic forces to inertial forces. Based on these dimensionless numbers, the analytical solution of the startup of a viscoelastic channel flow is derived following the work of Waters et al. as well as Xue et al. [208, 248]. The transient velocity profile can be written as

$$v_x^a(t) = \frac{F\varrho}{12\eta}\left[6y^*\left(y^* - 1\right) + \sum_{n=1}^{\infty}\frac{48}{N^3}\sin(Ny^*) \times \exp\left(-\alpha_n\frac{1}{2}\frac{t}{\lambda_1}\right)G(t)\right] \tag{4.34}$$

where $y^* = y/H$, $N = (2n + 1)\pi$ and $El = Wi/Re$. In addition, the following abbreviations are used

$$\alpha_n = 1 + \frac{1-\beta}{4}N^2\,El, \qquad \beta_n = \sqrt{\alpha_n\alpha_n - El\,N^2}, \qquad \gamma_n = 1 - \frac{1+\beta}{4}\,El\,N^2.$$

and the transient expression $G(t)$ is evaluated as

$$G(t) = \begin{cases} \cos(\frac{\beta_n}{2}\frac{t}{\lambda_1}) + \frac{\gamma_n}{\beta_n}\sin(\frac{\beta_n}{2}\frac{t}{\lambda_1}) & \text{for } \beta_n > 1 - 4(\sqrt{El\,N^2} - 1)/(El\,N^2) \\ \cosh(\frac{\beta_n}{2}\frac{t}{\lambda_1}) + \frac{\gamma_n}{\beta_n}\sinh(\frac{\beta_n}{2}\frac{t}{\lambda_1}) & \text{for } \beta_n \leq 1 - 4(\sqrt{El\,N^2} - 1)/(El\,N^2) \end{cases}$$

Poiseuille flow of an Upper-Convected Maxwell Model

As stated above, the degree of viscoelasticity of the Oldroyd-B fluid can be controlled by the retardation ratio β. In the first part of the following section, the parameter β is set to zero. Therewith, the diffusive terms in the Oldroyd-B fluid model are removed and only the elastic stress tensor $\tau_e^{\alpha\beta}$ contributes to the deviatoric stress tensor $\tau^{\alpha\beta}$ depicted in equation (4.29). Thereby, the corresponding momentum equation changes from a parabolic partial differential equation for the full Oldroyd-B fluid to an hyperbolic partial differential equation for the UCM model. As discussed by Xue *et al.* in great detail, the hyperbolic nature of the UCM fluid flow raises exceptional difficulties for transient numerical simulations [208]. In the absence of dissipative effects in the momentum equation, small subgrid fluctuations grow over time, eventually resulting in oscillations and divergent results. For this reason, several stabilization techniques are developed for grid-based finite volumes and elements schemes, such as the *elastic viscous stress split*, *both sides diffusion* and *adaptive viscous split stress* methods [208]. These numerical instabilities arise from the strongly advective character of the momentum balance. Ellero and Tanner showed, that Lagrangian particle methods, such as SPH, are better suited for describing the transient flow of a UCM fluid, since the advective derivative is already included in the Lagrangian derivative as shown in equation (3.2) [106]. However, the fluctuations are not completely removed according to Ellero and Tanner, but stay bounded at least for moderate Weissenberg numbers ($Wi < 1$).

In the figures 4.30 and 4.31 the transient velocity profiles over the channel height obtained with the NSPH as well as NCSPH approach are shown together with the respective analytical solution. The regarded flow is characterized by a very low Reynolds number in the range of $Re = 8.2 \cdot 10^{-10}$ due to the high dynamic fluid viscosity. This results also in a very small Weissenberg number in the range of $Wi = 1.6 \cdot 10^{-9}$. But the dimensionless Elastic number El is relatively high with an approximate value of $El = 2$, which is characteristic for a flow of a polymer melt [249]. For comparison, the velocity profiles of both investigated SPH approaches are shown, with the NSPH particles marked by a red square $\square$ in the lower half of the channel, while the values obtained with the NCSPH approach are depicted by a blue circle $\bigcirc$ in the upper half. The respective analytic velocity profiles are sketched by the solid black line. In figure 4.30, the startup behavior of the UCM fluid is shown at the time instants $t = 0.01\,s, 0.02\,s, 0.03\,s, 0.04\,s$ and $0.05\,s$. Both SPH solutions are in good agreement with the analytical one. Yet, the NCSPH approach is able to resolve the discontinuity in the velocity profile more accurately. This can be explained by a smaller smoothing length to particle size ratio of $h/L_0 = 1.05$. For a stable simulation, the smoothing length to particle size ratio of the NSPH method has to be set to $h/L_0 = 1.55$ and the velocity gradient is so smeared by the larger extent of the interpolation domain. However, the sharp velocity gradient can be resolved more sharply by refining the discretization, while keeping the smoothing length to particle size ratio constant. This is also demonstrated in figure 4.30 with the velocity profile at the time step $t = 0.05\,s$ marked by the green crosses ($\times$). By refining the resolution from 40 particles distributed over the channel height to 80 particles, the NCSPH simulation converges to the analytical solution. At around $t \approx 0.05\,s$, the velocity profile shows its larges deflection and the viscoelastic nature decelerates the UCM fluid. The further temporal evolution of the velocity profiles of the UCM fluid over the channel height is shown in figure 4.31 at the time points $t = 0.05\,s, 0.07\,s, 0.09\,s, 0.11\,s, 0.13\,s$ and $0.15\,s$. Again, good agreement with the analytical solution is achieved by both methods, while increasing deviations from the analytical solution are observed for the NSPH approach in boundary near regions. This is especially prominent at time instant $t = 0.11\,s$ as shown in the detail in figure 4.31. As one can see, the no velocity slip boundary condition at the fluid wall interface is not ensured with increasing time. Though, the overall agreement between the NSPH simulation result and the analytical solution is not drastically deteriorated by the boundary effects as shown by the time evolution of the centerline particle velocity shown in figure 4.32. In the details, a time delay between the analytical center axis velocity and both SPH based solutions is observed. Nevertheless, the deviations do not accumulate over time and the steady state velocity value is reached with high accuracy.

As stated at the beginning of the section, the simulation of viscoelastic flow characterized by high Weissenberg numbers as well as Elasticity numbers is challenging for Eulerian methods [196]. Ellero and Tanner ascribe SPH the ability to simulate high Weissenberg number flow, while only showing simulation results characterized by Elasticity numbers around $El = 20$ [106]. Thence, simulations with comparable and even higher Elastic numbers are conducted with the presented methods in order to find their respective limitations. For the next simulation, the time constant of relaxation λ_1 of the UCM fluid model is increased to $\lambda_1 = 1\,s$, which results in an Elasticity number of $El = 20$. In figure 4.33, the temporal evolution of the velocity of the center axis particle is shown.

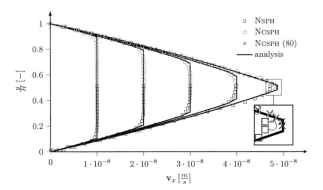

Figure 4.30: Startup of a viscoelastic UCM fluid Poiseuille flow with $El = 2$ over the dimensionless channel height. The velocity profile obtained by the NCSPH method are marked with $\bigcirc$, the one of the NSPH approach with $\square$. The shown profiles are extracted at $t = 0.01\,s, 0.02\,s, 0.03\,s, 0.04\,s$ and $0.05\,s$. At time step $0.05\,s$, the velocity profile obtained using the NCSPH approach with a higher particle resolution over the channel height is depicted with $\times$.

In the beginning of the simulation, the temporal evolution of the center axis velocity deviates slightly from the analytical solution as it has been observed previously for fluids characterized by $El = 2$. The deviations are especially pronounced at the reversal points as shown in the detail in figure 4.33. However, before reaching the steady-state velocity profile, the NSPH solution diverges at approximately $t \approx 4.99\,s$. This behavior can be attributed to spurious boundary effects at the solid wall, which lead to errors in the evaluation of the stress tensor. These errors accumulate and lead in terms to oscillations, which grow rapidly over time and stop the simulation. In contrast, a stable simulation is possible using the NCSPH approach, with the steady state velocity being in very good agreement with the analytical one. But, the small time lag between the NCSPH and analytical solution is further increased in the current case. This can be explained by the increased velocity gradient between the wall and the discontinuity in the velocity field with increased elasticity. To improve the resolution of the steep velocity gradient, the particle number needs to be increased as shown in the paragraph above.

The elastic character of the UCM fluid can also be increased by increasing the fluid

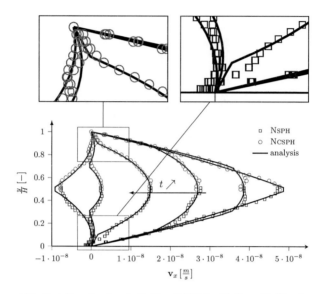

Figure 4.31: Evolution of a viscoelastic UCM fluid Poiseuille flow with $El = 2$ over the dimensionless channel height. The velocity profiles obtained by the NCSPH method are marked with $\bigcirc$, while the NSPH based results are indicated by $\square$. The shown profiles are extracted at $t = 0.05\,s, 0.07\,s, 0.09\,s, 0.11\,s, 0.13\,s$ and $0.15\,s$.

viscosity, while keeping the other process parameters constant. Therewith, the flow is also characterized by a higher Elasticity number, since the Reynolds number is decreased, while the Weissenberg number is kept constant. The respective temporal evolution of the center axis velocity of a UCM fluid with an increased viscosity η is shown in figure 4.34. Similar to the simulation shown above, the NSPH approach diverges abruptly after several time steps before reaching the steady state velocity. In contrast, the NCSPH based approach is capable of accurately simulating the temporal evolution. And as shown in the detail of figure 4.34, almost no time delay between the analytical and SPH based solution is observed in this case.

As announced above, the applicability of the presented NCSPH approach to simulate flows with high Elasticity numbers will be investigated. Therefore, the time constant of relaxation is further increased to $\lambda_1 = 10\,s$, which corresponds to an Elasticity number

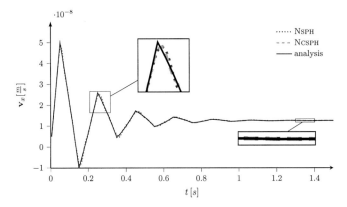

Figure 4.32: Evolution of the velocity $v_x(t)$ for a UCM fluid particle in the center of the Poiseuille channel after start-up until $t = 1.5\,s$. The flow conditions are characterized by $El = 2$.

of $El = 200$, and the temporal evolution of the corresponding horizontal velocity of the center axis is shown in figure 4.35. Satisfactory agreement between the theoretical evolution and NCSPH based simulation results is achieved. As expected, the NSPH based simulation did not converge and is consequently not suited for the simulation of viscoelastic materials with high Elasticity numbers. For the accurate and stable simulation of flows with high elasticity, the NCSPH based approach is the method of choice, since no stabilization techniques have to be used. Since only simulations with low Reynolds numbers have been conducted in the present work, the questions remains if the NCSPH approach is better suited for the simulation of high Weissenberg number flows. However, it has been demonstrated, that the problems in treating the no-slip boundary condition are effectively reduced. Due to its simple treatment of the no-slip velocity condition at fluid – solid interfaces, the NCSPH approach can be in principle also applied to flow in more complex geometries without modification. In the next paragraph, the NSPH and NCSPH approach are applied to simulate an Oldroyd-B fluid, with which the behavior of polymer melts and solutions can be simulated at least on a qualitative basis as stated initially [199].

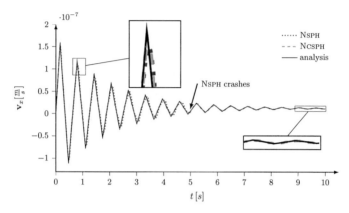

Figure 4.33: Evolution of the velocity $v_x(t)$ for a UCM fluid particle in the center of the Poiseuille channel after start-up until $t = 10\,s$. The flow conditions are characterized by $El = 20$.

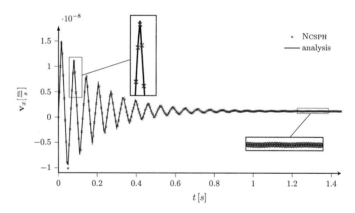

Figure 4.34: Evolution of the velocity $v_x(t)$ for a UCM fluid particle in the center of the Poiseuille channel after start-up until $t = 1.5\,s$. The flow conditions are characterized by $El = 20$.

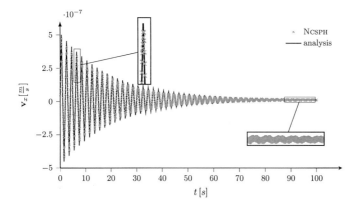

Figure 4.35: Evolution of the velocity $\boldsymbol{v}_x(t)$ for a UCM fluid particle in the center of the Poiseuille channel after start-up until $t = 100\,s$. The flow conditions are characterized by $El = 200$.

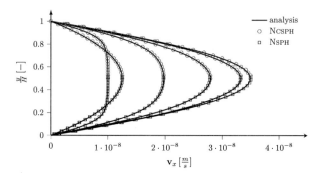

Figure 4.36: Evolution of a viscoelastic Oldroyd-B fluid Poiseuille flow with $El = 2$ over the dimensionless channel height. The velocity profiles obtained by the NCSPH method are marked with $\bigcirc$, while the NSPH based results are indicated by $\square$. The shown profiles are extracted at $t = 0.01\,s, 0.02\,s, 0.03\,s, 0.04\,s, 0.06\,s$ and $1.0\,s$.

Poiseuille flow of an Oldroyd-B fluid

In the previous section, the applicability of the derived SPH based approaches for simulating purely elastic fluids, like the UCM fluid, has been investigated. In the following section, the developed approaches are applied to fluid flow described by the Oldroyd-B constitutive equation. Due to its parabolic nature, the resulting momentum equation is easier to handle in a numerical point of view. However, the Oldroyd-B fluid model is interesting from an application point of view. By proper parameterization, the model is capable of describing the dynamic behavior of polymer melts and polymer solutions at least on a qualitative basis (Oldroyd 6-constant model) [199]. In figure 4.36, the simulation results obtained by the NSPH and NCSPH approach are compared to the respective analytical solution for a retardation ratio of $\beta = 0.9$. In both cases, very good agreement between the SPH based solution and the theoretical results is achieved. Therewith, the applicability of the developed SPH based approaches to model viscous, viscoplastic as well as viscoelastic fluids has been demonstrated. In the subsequent section, their applicability towards modeling the behavior of elastic solids is investigated.

Parameter		Value
initial particle distance	L_0	$0.025\,ul$
maximum time step	Δt	$10^{-6}\,s$
cutoff radius NCSPH	r_{cut}	$2.1\,L_0$
cutoff radius ISPH\NSPH	r_{cut}	$3.1\,L_0$
channel height	H	$40\,L_0$
initial fluid density	ϱ^0	$1\,\frac{kg}{L_0^3}$
dynamic viscosity	η	$10-100\,\frac{kg}{ul\,s}$
relaxation constant	λ_1	$0.1-10\,s$
retardation constant	λ_2	$0-0.01\,s$

Table 4.14: Computational details of the simulation of the Poiseuille flow of a viscoelastic fluid. All measures of length are given in chosen units abbreviated by ul as unit length.

4.2.4 Elastic materials

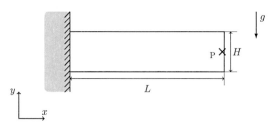

Figure 4.37: Computational domain of the elastic beam bending test case. The cantilever beam has a fixed bearing on the left side and bends under the influence of gravity. The position of the reference particle P is marked by the cross.

So far, the applicability of the derived SPH approaches towards different viscous material behavior has been studied and verified. In order to simulate the morphogenesis of open-porous materials, the treatment of solid inclusions in viscous substrates and their mutual interaction is an important aspect. This is especially true for the regarded application process, as introduced in section 1.2, where preferential regions of deformation and viscous flow develop due to the anisotropy introduced by the solid inclusion. As a further example, the formation of porous structures during spray drying of polymer suspensions may serve [220]. Depending on the application, several approaches are possible to take solid inclusions into account. For example, the solids can be treated as rigid bodies, if the deformation of the solid is negligible or does not influence the overall result. An example would be the circulation of a solid cylinder by a viscoplastic fluid or low Reynolds fluid flow through porous media [250–252, 252, 253, 253–257]. In the present work, a more versatile approach is used, which attributes a certain material strength to the solid bodies. Hence, a possible influence of the solid deformation on the resulting structure and morphology is included in the simulation. In the following section, the discretization of the stress tensor for elastic materials is presented. In principle, two different formulations for modeling linear elastodynamics with particle methods are used in the literature. In the so called displacement based approach, the Navier-Lamé equations are directly solved [114, 142]. It has been shown, that this approach yields the most accurate and stable results if using the corrected second order spatial derivative shown in equation E.1. As stated by Chen *et al.* as well as Belytschko *et al.*, this is due to the integrability and completeness of the corrected second order discretization scheme [80, 142]. However, the classical stress based approach, as shown in section 3.1, has been used in several works in the context of SPH and is also used in the present work [78, 155, 221]. Therewith, the coupling of fluid and solid dynamics to a dynamic multiphysic simulation is simplified. In contrast to the former method, where the stress is calculated from the particle displacement, the stress is obtained from the strain rate tensor as it will be shown in the following.

As constitutive equation for the isotropic solid Hooke's law is used. So, the overall stress tensor $\boldsymbol{\sigma}$ is written as [159]

$$\sigma^{\alpha\beta} = 2\mu\epsilon^{\alpha\beta} + \lambda\delta^{\alpha\beta}\epsilon^{\gamma\gamma}, \qquad (4.35)$$

where ϵ mimics the strain tensor and λ represents the first Lamé constant and μ the second Lamé constant or shear modulus. The rate of change of the stress tensor is then given by

$$\frac{D\,\sigma^{\alpha\beta}}{D\,t} = 2\mu\dot{\epsilon}^{\alpha\beta} + \delta^{\alpha\beta}\lambda\dot{\epsilon}^{\gamma\gamma} + \sigma^{\alpha\gamma}\Omega^{\gamma\beta} + \Omega^{\alpha\gamma}\sigma^{\gamma\beta}, \qquad (4.36)$$

where summation is implied over repeated indices. The rotation or vorticity tensor $\boldsymbol{\Omega}^{\alpha\beta}$ of the corotational derivative is given by

$$\Omega^{\alpha\beta} = \frac{1}{2}\left(\frac{\partial v^{\alpha}}{\partial x^{\beta}} - \frac{\partial v^{\beta}}{\partial x^{\alpha}}\right). \qquad (4.37)$$

By using the corotational derivative (Jaumann derivative), the applicability of Hooke's law is extended to cope with finite displacements [199]. Under these finite displacement conditions, the original equation is not material frame indifferent. In this way, the material response is depending on the rotation of the material and observer in an unphysical way [221]. And this phenomenon is corrected by the corotational derivative. In addition, one has to stress, that the strain rate tensor $\dot{\epsilon}$ is defined slightly different as in the previous cases to give

$$\dot{\epsilon}^{\alpha\beta} = \frac{1}{2}\left(\frac{\partial v^{\alpha}}{\partial x^{\beta}} + \frac{\partial v^{\beta}}{\partial x^{\alpha}}\right). \qquad (4.38)$$

With known Poisson ratio ν and Young's modulus E, the shear modulus is estimated to

$$\mu = \frac{E}{2(1+\nu)} \qquad (4.39)$$

and the first Lamé constant is estimated as

$$\lambda = \frac{\nu E}{(1 + \nu)(1 - 2\nu)}. \tag{4.40}$$

In the current two-dimensional, respectively plane-stress case, the first Lamé constant has to be further modified to

$$\overline{\lambda} = \frac{2\lambda\nu}{(\lambda + 2\nu)} \tag{4.41}$$

and is used in equation (4.36) for the test case at hand.

For testing the applicability of the presented SPH approaches for the simulation of elastic material deformation, the response of a cantilever beam by sudden exposure to a body force is investigated. The geometry of the test case is shown in figure 4.37, with a length to height ratio of $L/H = 4$. The left edge of the beam is fixed and a body force of $g = 9.81 \, ul/s^2$ is applied to the beam immediately after the beginning of the simulation. The beam is discretized by 40×11 particles distributed on a regular lattice. The smoothing length to initial particle spacing ratio is set to $h/L_0 = 1.05$, which results in a cutoff radius of $r_{cut} = 2.1 \, L_0$ for the cubic spline kernel. The elastic material is characterized by a Young's modulus of $E = 2.1 \cdot 10^4 \, kg/ul \, s^2$ and a Poisson ratio of $\nu = 0.3$. Further computational details of the simulation are summarized in table 4.15. The fixed boundary at which the Dirichlet boundary condition for the displacement $\boldsymbol{u}$

$$\boldsymbol{u} = 0 \tag{4.42}$$

is specified, is represented by one layer of 11 of fixed boundary particles in the NCSPH approach and by two additional layers with 2×11 fixed particles using the NSPH approach. Since in the latter case, the boundary has to be reconstructed if using a non-corrected gradient approximation as shown in section 2.5. At the free boundary particles, the stress-free boundary condition

$$\boldsymbol{\sigma} \cdot \hat{\boldsymbol{n}} = 0 \tag{4.43}$$

is applied in both approaches, with $\hat{\boldsymbol{n}}$ being the outward pointing unit normal vector. As stated in section 3.2.2.2 and shown in section 2.1.3, the stress-free boundary condition is approximately, but intrinsically satisfied using the NSPH approach. In contrast, the boundary condition at the free surface has to be enforced using the corrected NCSPH approach. In two dimensions, the normal stress $\sigma_{n,n}$ and the shear stress components

$\sigma_{n,t}$ and $\sigma_{t,n}$ at the free surface must vanish. Therefore, the stress tensor $\boldsymbol{\sigma}$ (as well as the velocity gradient tensor) is rotated in the normal plane spanned by the outward pointing normal vector $\hat{n}$ and the tangent vector at the surface $\hat{t}$. In the new coordinate plane, the respective tensor components are zeroed out and the tensors are rotated back into the Cartesian coordinate system as stated in section 3.2.2.1.

The time histories of the displacement of the rightmost particle in the centerline of the beam, indicated in figure 4.37 by the black cross, are shown in figure 4.38 for the NSPH ($\square$) and NCSPH ($\bigcirc$) approach. In addition, a finite element based solution (Ansys$^{\circledR}$ Structural) is depicted by the solid black line as a reference. The reference solution is resembled very well by both SPH solutions. Though, the normalized corrected approach yields slightly more accurate results as can be seen in the details of figure 4.38. In order to study the influence of a larger displacement on the accuracy, the local gravity is increased by a factor of ten and the time history of the resulting displacement is shown in figure 4.39. No prominent degradation in accuracy is observed for both methods under these small displacements.

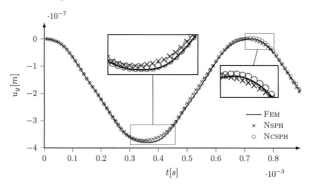

Figure 4.38: Time history of the horizontal displacement u_y of the center particle of the far right edge of the beam under the local gravity g. The trajectories of the regarded particles ($\bigcirc$ NCSPH, $\times$ NSPH) begin from a position at rest and are compared to the finite element based solution ($-$).

189

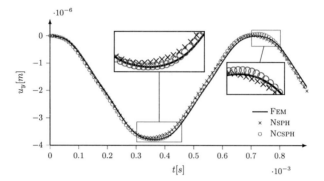

Figure 4.39: Time history of the horizontal displacement u_y of the center particle of the far right edge of the beam under an increased local gravity of $10\,g$. The trajectories of the regarded particles ($\bigcirc$ NCSPH, $\times$ NSPH) begin from a position at rest and are compared to the finite element based solution ($-$).

Parameter		Value
initial particle distance	L_0	$10^{-4}\,ul$
maximum time step	Δt	$10^{-6}\,s$
cutoff radius	r_{cut}	$2.1\,L_0$
beam height	H	$11\,L_0$
beam length	L	$40\,L_0$
surface boundary	β^S	0.95
initial density	ϱ^0	$1\,\frac{kg}{ul^3}$
Poisson number	ν	0.3
Young's Modulus	E	$2.1 \cdot 10^4\,\frac{kg}{ul\,s^2}$

Table 4.15: Computational details of the simulation of the bending of a elastic cantilever beam. All measures of length are given in chosen units abbreviated by ul as unit length.

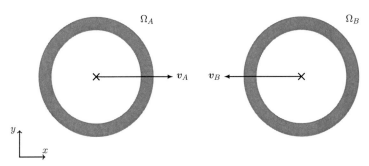

Figure 4.40: Initial configuration of the elastic rubber rings in the collision test case.

Rubber ring collision

So far, both presented NSPH and NCSPH approaches proved to be capable to treat elastic deformations in an accurate and stable way. However, for simulations involving rapid material deformations, it is known, that SPH based approaches are becoming increasingly unstable [78]. If solving the conservation equations together with a purely elastic constitutive equation as Hooke's law, unphysical numerical oscillations occur after shocks and during their propagation as shock waves, e.g. after collision of two solid bodies. This is not only true for a SPH based discretization, but common for many numerical approaches as stated by Libersky and coworker, since dissipative contributions are neglected in the numerical scheme [221]. Hence, the concept of *artificial viscosity*, invented by Von Neumann and Richtmayer in the context of the finite difference method and transferred to SPH by Monaghan and Gingold, is commonly used in the literature to stabilize the numerical computation by smoothing possible shocks [236, 258]. In the literature, this stabilization technique is also applied within the Normalized Corrected SPH framework [144]. In doing so, the artificial viscosity is commonly included as an additional viscous pressure term in the momentum balance and estimated by the following expression [236]

$$\Pi_{ij} = \begin{cases} \dfrac{-\alpha c^0 \mu_{ij} + \beta \mu_{ij}^2}{\bar{\varrho}_{ij}} & \text{for} (v_j - v_i) \cdot (x_j - x_i) < 0 \\ 0 & \text{otherwise} \end{cases} , \qquad (4.44)$$

where

$$\mu_{ij} = \frac{h(v_j - v_i) \cdot (x_j - x_i)}{|x_j - x_i| + \kappa h^2} \qquad (4.45)$$

and $\bar{\varrho}_{ij}$ being the mean density and c^0 the reference speed of sound. The reference speed of sound is calculated with known Young's modulus E and Poisson ratio ν as

$$c_0 = \sqrt{\frac{E}{3(1+\nu)\varrho}}. \qquad (4.46)$$

The additional parameters α and β are commonly chosen of order unity and $\kappa = 0.1$ [221, 259, 260]. As stated by Chen and coworker, the first term in equation (4.44) acts like an additional bulk and shear viscosity term, while the second is equivalent to the Von Neumann-Richtmayer viscosity [144]. Besides the physical motivation of the presented artificial approach, the fitting parameters α, β and κ are introduced in the numerical scheme. This is the main reason for disregarding the concept of artificial viscosity in the presented work. In order to eliminate the stability problems, a simple additional smoothing of the velocity field is sufficient. The used approach is based on the corrected kernel estimate (equation (2.64))

$$\widetilde{v}_i = v_i + \alpha_S \left(\frac{\sum_{j\neq i} V_j v_j w_{ij}}{\sum_{j\neq i} V_j w_{ij}} - v_i \right), \qquad (4.47)$$

which can be seen as a generalization of the conservative smoothing approach introduced by Guenther *et al.* and refined and applied by several authors [155, 261–264]. The numerical parameter α_S is known as the conservative smoothing coefficient with $\alpha_S = [0, 0.5]$. With the smoothed velocity field $\widetilde{v}$ the respective strain rate and vorticity (equations (4.28) ,(4.37)) are evaluated. The smoothing coefficient should be chosen as small as possible to minimize the influence of the additional smoothing. One has to stress, that the additional smoothing of the velocity field resembles some similarities with the *Extended* SPH approach, which is abbreviated by XSPH. XSPH has been introduced by Monaghan to improve the stability of SPH simulations by helping to maintain an orderly and regular particle distribution [164]. This is especially true for weakly compressible SPH simulations [67, 213] and simulations with elastic material models involving tension, where the XSPH approach is used in conjunction with the artificial viscosity and artificial stress [78]. With the XSPH approach, the particle velocity v_i is replaced by an averaged velocity $\hat{v}_i$ for the calculation of the particle positions only. The averaged velocity $\hat{v}_i$ is obtained by taking the velocities of the neighboring particles into consideration and is defined as

$$\hat{v}_i = v_i + \epsilon \sum_j \frac{2\,m_j}{\varrho_i + \varrho_j} (v_j - v_i) w_{ij}, \qquad (4.48)$$

where the parameter ϵ is generally set to $\epsilon = 0.5$ [164]. For consistency reasons, $\hat{v}_i$ has also to be used if the density is calculated via equation (3.33) [164]. Similar to the latter approach, the modified velocity is used to calculate the strain rate and vorticity tensor if the behavior of solid materials is modeled.

By using the simpler conservative smoothing process, shown in equation (4.47), possible tensile instabilities are suppressed. For validation, the collision of two elastic rubber rings as sketched in figure 4.40 is investigated. The example is considered as a severe test in the literature for the applicability of the algorithm to handle compression and tension of thin structures [78, 213, 265, 266]. For describing the contact of two self-contained bodies, the frictionless contact algorithm of Vignjevic et al. is inherited. Details regarding the derivation of the respective algorithm can be found in [267, 268]. In brief, the contact force between the approaching bodies is estimated by the contact potential, which is defined as

$$\Phi_i^c = \sum_{j \notin \Omega_i} \frac{m_j}{\varrho_j} K \left(\frac{w^C(r_{ij})}{w^C(\overline{r})} \right)^n, \tag{4.49}$$

with $w^C(r_{ij})$ being the contact weight function evaluated with the distance between the interacting particles and $w^C(\overline{r})$ being the contact weight function evaluated with the average particle spacing in the neighborhood of particle i. In the expression, the summation is extended over the neighboring particles of the opposing body only and K is a contact stiffness penalty parameter and the exponential parameter n determines the slope of the potential. In principle, any weight function w^C can be used for describing the contact potential. However, the following weight function

$$w^C(r_{ij}) = r_{cut}/r - 1 \tag{4.50}$$

is used in the present work, since the value of the function tends to infinity for the particle distance going to zero. The respective function is taken from the Moving Particle Semi-implicit method, where it is used as kernel function in the kernel interpolation expressions [193]. The contact force of the regarded particle $\boldsymbol{f}_i^c$ is obtained from the spatial gradient of the potential

$$\boldsymbol{f}_i^c = \sum_{j \notin \Omega_i} \frac{m_j}{\varrho_j} \frac{m_i}{\varrho_i} K \, n \, \frac{w(r_{ij})^{n-1}}{w(\overline{r})^n} \nabla_j w(r)_{ij}. \tag{4.51}$$

Thereby, the summation is only extended over the particles of the opposing body. The influence radius for the contact force is set to $3.1\,L_0$, with L_0 being the initial particle spacing. The contact stiffness penalty parameter K is set to unity and the exponential parameter to $n = 4$ as suggested by the inventors [267]. The resulting contact force is included in the momentum equation as an additional body force. For the simulation of the collision, rubber rings with an initial inner radius of $3\,ul$ and an outer radius of $3.5\,ul$ are chosen. Each ring is modeled by 1016 particles placed on a square lattice, since the alignment on a square lattice is known to be more sensitive to tensile instabilities [26]. For this reason, the discretized rings have a rough surface and steplike appearance. The initial particle spacing is $L_0 = 0.1\,ul$ and a smoothing length to initial particle distance ratio of $h/L_0 = 1.05$ is used. Each ring moves with an initial velocity of $10\,ul/s$, resulting in a relative velocity of $20\,ul/s$. Further computational details are summarized in table 4.16.

In figure 4.41, the chronology of the collision simulated with the NCSPH approach without any stabilization, e.g. by conservative smoothing ($\alpha_S = 0$) is shown. In the left picture, the undeformed particle configuration right before the impact is shown. On the right at time step $t = 2.7 \cdot 10^{-2}\,s$ the collision of the rubber rings is shown. Both rings fracture almost immediately upon collision due to tensile instability. Since this observation also holds for the NSPH approach, both methods are not capable of describing rapid and large elastic material displacement. Hence, the simulation is repeated with an conservative smoothing coefficient of $\alpha_S = 0.3$. The respective chronology is shown in figure 4.42 for the NCSPH approach. After collision, the rings deform and the maximal compression is reached at about $t = 35 \cdot 10^{-2}\,s$. After that, the rings bounce off again. As one can see, tensile instabilities are successfully suppressed by the conservative smoothing of the velocity. No velocity fluctuations and particle clumping are observed during the computation, which is considered as an indicator for developing instabilities [265, 266]. So it is demonstrated, that the additional smoothing of the velocity field is necessary for the dynamic simulation of elastic materials under rapid as well as large displacements. However, the smoothing needs to be as small as possible to minimize the artificial dissipation effect. Since sharp velocity gradients are smeared out, local strain and stress is reduced and the overall dynamic of the elastic body is changed. After collision, the rings are not retaining the initial absolute velocity again and do not continue to oscillate freely in space. For example the velocity of each

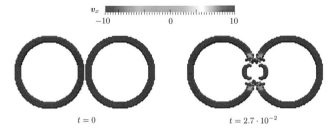

Figure 4.41: Snapshots of the initial position and velocity of the colliding elastic rings and fracture upon collision simulated with the NCSPH method without using conservative smoothing ($\alpha_S = 0$).

ring comes up to an absolute value of $|v| = 3.1$ at $t = 2.2\,s$ only. In addition, the rings fracture after $t = 2.6$ before retaining their initial configuration. Accordingly, the relative low absolute velocity after the impact gives rise to the fact, that a considerable amount of dissipation is added by using the conservative smoothing with $\alpha_S = 0.3$, even though the rings are still considerably deformed. And a further reduction of α_S is not an option, since a stable simulation cannot be guaranteed under these conditions. To conclude, the presented approach is not capable of describing elastic material behavior under rapid and large deformation in a quantitative manner. If the simulation of elastic materials under these conditions are desired, either an optimization of the conservative smoothing approach or the application of the artificial viscosity and artificial stress concept is necessary as introduced by Gray and coworkers [78]. As a further option, an alternative formulation of the momentum balance based on the particle displacement as discussed in the first part of this section should be used. The displacement based approach was found to be more stable and insensitive to tensile stress [269]. However, for the desired application, fast and large deformations of purely elastic materials are excluded and the focus is put on a stable simulation. Nevertheless, it has been proven, that quantitative simulation results are obtained for small deformations as shown by the deformation of the elastic cantilever beam.

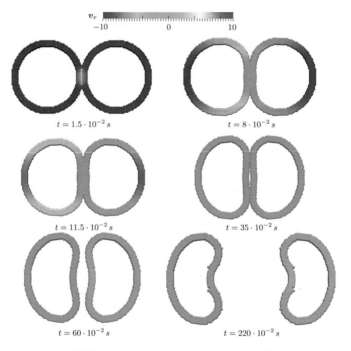

Figure 4.42: Snapshots of the collision of two elastic rings simulated by the NCSPH method using a conservative smoothing coefficient of $\alpha_S = 0.3$. Positions and velocities of the colliding elastic rings are depicted at $t = 0.015\,s, 0.08\,s, 0.115\,s, 0.35\,s, 0.6\,s$ and $2.2\,s$.

Parameter		Value
initial particle distance	L_0	$0.1\,ul$
maximum time step	Δt	$10^{-6}\,s$
cutoff radius	r_{cut}	$2.1\,L_0$
outer ring diameter	R_i	$3\,ul$
inner ring diameter	R_a	$3.5\,ul$
absolute velocity	$|v_x|$	$10\,ul/s$
surface boundary	β^S	0.95
initial density	ϱ^0	$1\,\frac{kg}{ul^3}$
Poisson number	ν	0.3
Young's Modulus	E	$10^4\,\frac{kg}{ul\,s^2}$

Table 4.16: Computational details of the simulation of the collision of two elastic ruber rings. All measures of length are given in chosen units abbreviated by ul as unit length.

4.2.5 Interim conclusion

In the previous section, the applicability of the developed Sph approaches for modeling various materials has been investigated. Thereby, no significant increase in accuracy has been observed by using renormalized or renormalized corrected Sph approaches in comparison to the conventional approach. This is quite natural, since the computational domains of the used test cases are not truncated by free boundaries. However, one has to stress, that a renormalization and correction of the kernel function strongly improves the stability of the simulation. This is especially true for viscoelastic materials, where the reconstruction of the time-dependent stress tensor for the solid wall particles embodies a difficult problem. Therewith, the Ncsph approach with its natural treatment of the no-slip velocity boundary condition is the method of choice and yields accurate as well as stable simulation results. Furthermore, it has to be noted, that rapid deformations of elastic materials are only possible if additional smoothing techniques are applied.

4.3 Validation and verification of the multiphase algorithm

In the latter section, the suitability of the developed SPH based approaches to single phase processes has been demonstrated. In general, the Normalized Corrected SPH (NCSPH) approach proved to be especially suited for the simulation of structure forming processes. Due to the applied corrections, boundary conditions can be applied in a simple and consistent manner, making the NCSPH a versatile, robust and accurate simulation approach for problems with complex material behavior and free surfaces. So far, the presented SPH approaches succeed in correctly simulating single phase flows. However, most of the morphogenesis processes are governed by the interaction of phases with different aggregate states. In addition, interfacial forces, such as surface tension and wettability effects, are playing a crucial role in many chemical engineering applications. Though, the presence of interfaces and the associated change in physical quantities over the interface makes the simulation of multiphase systems more challenging. In the following section, the presented approaches are tested for their applicability to simulate the evolution of multiphase systems. The focus is laid on the treatment of multiphase systems consisting of incompressible and compressible phases, as they occur in the desired application of reaction-induced morphogenesis of open-porous materials. Therefore, the coupling of incompressible and compressible phases presented in section 3.2.1 is validated by means of the simple hydrostatic test in the following.

4.3.1 Hydrostatic pressure in a channel with a free surface

As stated above, the treatment of multiphase systems and especially of systems with large density difference between the respective phases is still a current challenge within the development of the SPH method [81, 82, 119, 123, 135, 169, 170, 211, 212, 236, 270]. As stated by Ott and Schnetter, this is especially true for systems with a density ratio between the respective phases exceeding one order of magnitude [271]. Due to the kernel interpolation, shown in section 2.1, only density gradients smaller than the gradient of the smoothing kernel are allowed [92, 135]. This limitation still persists, if using the corrected SPH approach, which is used in the present work and introduced in section 2.5 as reported by Liu and Li [52]. Since the kernel approximation and its derivative are based on a Taylor series expansion, the considered function needs to be sufficiently smooth. In order to accurately treat discontinuities, a further modification

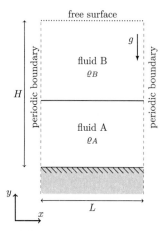

Figure 4.43: Computational domain of the hydrostatic test case. The channel is filled with two fluids A and B with different densities ϱ_A and ϱ_B.

to the corrected SPH approach has been proposed by Liu and coworkers, which is called Discontinuous SPH (DSPH) [272]. In that approach, an additional corrective term is introduced, which approximates the discontinuity. Yet, the computational costs are greatly increased, since the nearest neighbors of the discontinuity have to be tracked. Consequently, several approaches within the classical SPH method have been suggested to overcome the limitations of the kernel estimate. For example, Ritchie and coworker assume a constant particle averaged pressure across the kernel function and calculate the respective density subsequently from an equation of state [82]. Therewith, mass conservation is lost. Colagrossi *et al.* handles large density gradients, as stated in section 2.2, by modifying the first order spatial derivative to reduce the influence of large density gradients over the interface [81]. Supplemental, a sort of reinitialization or repacking of the particle has been applied and the density needs to be obtained from the non-conservative formulation shown in equation (3.33). That approach was refined by Grenier *et al.* in order to conserve the fluid mass [119]. Hu and coworkers derived particle-averaged spatial derivatives, in which the density of the neighboring particles is replaced by the particle number density [123]. That approach has been applied to the weakly compressible as well as truly incompressible SPH approach [169, 170, 273].

However, both approaches are not suited for problems involving free surfaces and the incompressible approach has only been applied to problems involving density ratios of three [169]. Recently, Monaghan and Rafiee reported, that all the previously described approaches are not necessary using the WCSPH approach under the constraint of a sufficient resolution of the periphery of the discontinuity [212]. Nevertheless, Monaghan and Rafiee used an artificial surface tension term for keeping the fluid phases separated. In the present work, the approach of Grenier *et al.* is advanced to treat multiphase systems, consisting of truly incompressible as well as compressible phases as presented in section 3.2. In order to avoid the summation of the density over the discontinuity, the density is only added up within the respective phase as shown in equation (3.40) and (3.94). Thus, the discontinuity in the density field is treated explicitly. And by calculating the density via summation over the neighboring particles, the mass is implicitly conserved. The connection between the phases is recovered by calculating the volume of each particle via the continuity equation as shown in equation (3.45) and (3.95). Since the velocity is continuous over the interface, the gradient expression based on the SPH kernel interpolation can be used.

In the following section, the multiphase approach presented in section 3.1 is validated by means of the simple hydrostatic test case already introduced in section 4.1.2 for single phase flow. In the present test case, two fluids, fluid A and fluid B, are layered in the channel as shown in figure 4.43. The computational domain is discretized by 30 particles over the channel height. The particles are distributed on a regular lattice. The cutoff radius is set to $r_{cut} = 2.1 L_0$, with L_0 being the initial particle spacing. All further computational details are summarized in table 4.17. In order to investigated the performance of the developed NCSPH approach for simulating multiphase flow, simulations with different density ratios ϱ_A/ϱ_B are performed and compared to the respective analytical solution. For each simulation, the density ratio ϱ_A/ϱ_B is increased by a factor of ten, starting with $\varrho_A/\varrho_B = 1$ and ending with $\varrho_A/\varrho_B = 10^4$. The steady state pressure profiles over the channel height are depicted in figure 4.44 on the left side in a semi-logarithmic form with increasing density ratios. On the right side of figure 4.44 the respective normalized root mean square errors, as defined in equation (4.16), are shown for the lower half of the computational domain. For a density ratio of $\varrho_A/\varrho_B = 1$, very good agreement between the analytical solution and the SPH based multiphase approach is achieved with an maximum relative error of less than 0.5 %. But the local

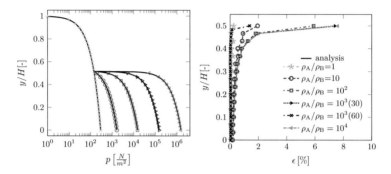

Figure 4.44: Left: Steady state pressure profiles of the compressible – incompressible multiphase NCSPH approach for various density ratios in comparison to the analytical solution in a semi-logarithmic representation. Right: Respective normalized root mean square error over the channel half-height.

error increases especially at the interphase with increasing density ratio ϱ_A/ϱ_B. For a density ratio of $\varrho_A/\varrho_B = 10$, the deviation from the analytical solution at the interface adds up to approximately 2 % and less than 6 % in case of a ratio of $\varrho_A/\varrho_B = 100$. For a density ratio of $\varrho_A/\varrho_B = 10^3$, being in the similar range as the water – air system, the relative error accumulates to approximately 8 %. With a further increase of the density ratio to $\varrho_A/\varrho_B = 10^4$, no signification increase in the local error is observed. Yet, the deterioration in accuracy can be effectively reduced by increasing the discretization resolution to 60 particles over the channel height. Therewith, very good agreement with the analytical solution is achieved and the maximum local normalized root mean square error is calculated to less than 1.2 %. Finally, one has to stress, that the simulation results obtained with the multiphase NSPH approach are slightly less accurate, since the interpolation domain of the Laplacian operator in the pressure Poisson equation is truncated by the interface. The impact of the truncation has been investigated in section 4.1.2.

Based on the presented results, the developed compressible – incompressible coupling scheme proved capable of simulating multiphase systems with large density ratios in a stable and accurate manner. However, for perfect agreement between the analytical and the SPH solution, a sufficient resolution of the interface is needed for large density ratios. This has also been recently reported by Monaghan and Rafiee [212].

Parameter		Value
initial particle distance	L_0	$1\,ul$
channel height	H	$30\,L_0,\ 60\,L_0$
cutoff radius	r_{cut}	$2.1\,L_0$
surface bound	β^S	0.95
initial density fluid A	ϱ^0	$1 - 10^4\,\frac{kg}{ul^3}$
initial density fluid B	ϱ^0	$1\,\frac{kg}{ul^3}$
maximum time step	Δt	$10^{-6}\,s$

Table 4.17: Computational details of the simulation of the hydrostatic test case with two immiscible fluids with different density ratios $\frac{\varrho_A}{\varrho_B}$. All measures of length are given in chosen units abbreviated by ul as unit length.

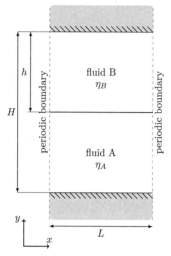

Figure 4.45: Computational domain of the two-dimensional Poiseuille flow test case with two layered immiscible and incompressible phases.

4.3.2 Multiphase Poiseuille flow of a viscous fluid

In the following section, the applicability of the current SPH models to multiphase flow with a high viscosity ratio between the phases is investigated. Therefore, the hydrostatic test case of layered fluids shown in figure 4.43 is evolved to the flow of two incompressible and immiscible Newtonian fluids along the channel centerline under the influence of a horizontal pressure gradient. The computational domain is similar to the single phase Poiseuille flow test cases. However, the two immiscible and incompressible fluids are layered in the channel as sketched in figure 4.45. For the simulation, the pressure gradient along the channel length is mimicked by a volume force acting on all particles.

In the test case, the nested gradient approach as well as the hybrid Laplacian formulation for modeling viscous effects is investigated. For the nested gradient approach, the divergence of the deviatoric stress tensor is directly calculated from the components of the stress tensor itself. The stress tensor is calculated by

$$\tau_i^{\alpha\beta} = \sum_j \frac{2\eta_i\eta_j}{\eta_i + \eta_j} V_j [(v_j^\alpha - v_i^\alpha)\widetilde{\nabla}_i^\beta \widehat{w}_{ij} + (v_j^\beta - v_i^\beta)\widetilde{\nabla}_i^\alpha \widehat{w}_{ij}], \qquad (4.52)$$

where the summation is extended over both phases. It has to be pointed out, that for the simulation of multiphase flow with different viscosities, it is absolutely necessary to use an inter-particle averaged viscosity. In the present work the harmonic mean of the fluid viscosities has been found to give the most accurate results for multiphase systems. The conventional approach of treating the viscosity independently of the velocity gradient results in oscillations at the interface, which deteriorate the accuracy and even crash the simulation. In the second approach, the divergence of the stress tensor is discretized using equation (4.14)

$$\nabla \cdot \boldsymbol{\tau} = (\nabla \cdot \eta\nabla)\, \boldsymbol{v} = \sum_j V_j \frac{4\,\eta_i\eta_j}{\eta_i + \eta_j} \frac{(\boldsymbol{x}_j - \boldsymbol{x}_i) \cdot \widetilde{\nabla}\widehat{w}_{ij}}{r_{ij}^2}(v_j - v_i), \qquad (4.53)$$

and the harmonic mean of the dynamic viscosities of the respective interacting particles η_i and η_j is used.

In the following, only the results obtained by the NCSPH approach are depicted, but the NSPH approach shows similar results. The density of both fluids is set to $\varrho = 1\,kg/ul^3$ and the viscosity ratio η_A/η_B between both phases is varied for each simulation. The cutoff radius for the kernel interpolation is kept at $r_{cut} = 2.1\,L_0$ and the cubic spline kernel is used. The maximum allowable time step is limited to $t = 10^{-4}\,s$. Further computational details are summarized in table 4.18. Since the flow considered is sufficiently slow and laminar, no instabilities at the fluid interface occur. Therewith, the interface remains at the channel centerline at all times and the steady state analytical solution for the considered flow can be derived as shown by Reis and Phillips to [274]

$$v^a = \begin{cases} \frac{Fh^2}{2\eta^A}\left(-(\frac{y}{h})^2 + \frac{y}{h}\frac{\eta^A - \eta^B}{\eta^A + \eta^B} + \frac{2\eta^A}{\eta^A + \eta^B}\right) & \text{for } -h \leq y \leq 0 \\ \frac{Fh^2}{2\eta^B}\left(-(\frac{y}{h})^2 + \frac{y}{h}\frac{\eta^A - \eta^B}{\eta^A + \eta^B} + \frac{2\eta^A}{\eta^A + \eta^b}\right) & \text{for } 0 < y \leq h, \end{cases} \qquad (4.54)$$

with h being the half plane of the channel with a value of $h = H/2$ and y the dimensionless height vector of the half plane.

In figure 4.46, the steady state velocity profiles are shown over the channel height for two different viscosity ratio. In the left plot, the solution is obtained by the nested gradient based formulation of the stress tensor (equation (4.52)), while the solution based on the Laplacian operator is shown on the right (equation (4.53)). All results are compared to the analytical solution, which are marked by the solid lines. For both investigated viscosity ratios $\frac{\eta^A}{\eta^B} = 10$ and $\frac{\eta^A}{\eta^B} = 100$, excellent agreement with the analytical solution is obtained. The normalized root mean square errors, as introduced in equation (4.17), are far below 1% for all combinations of the SPH versions, discretization schemes and viscosity ratios. Thus, it is assured, that the current models are capable of simulating multiphase flow with large viscosity ratios in a stable manner and with high accuracy. Together with the accurate description of the interaction of compressible gas and incompressible fluid phases as demonstrated in the previous section 4.3.1, the buoyancy driven rise of an immersed air bubble in water is investigated in the following section.

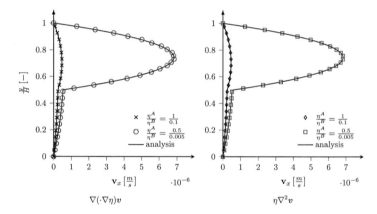

Figure 4.46: Multiphase Poiseuille flow of two layered fluids with different viscosities and discretization schemes for the deviatoric stress tensor. Steady state velocity profiles over the channel height for the viscosity ratio $\frac{\eta^A}{\eta^B} = 10$ and $\frac{\eta^A}{\eta^B} = 100$ using the NCSPH approach in comparison to the theoretical solution.

Parameter		Value
initial particle distance	L_0	$0.025\,ul$
cutoff radius	r_{cut}	$2.1\,L_0$
initial density fluid A	ϱ^0	$1\,\frac{kg}{ul^3}$
initial density fluid B	ϱ^0	$1\,\frac{kg}{ul^3}$
dynamic viscosity fluid A	η_A	$0.5 - 1\,\frac{kg}{ul\,s}$
dynamic viscosity fluid B	η_B	$0.005 - 0.1\,\frac{kg}{ul\,s}$
volume force per particle	F_i	$= 10^{-6}\,\frac{N}{ul^3}$
maximum time step	Δt	$10^{-4}\,s$

Table 4.18: Computational details of the simulation of the Poiseuille flow of two immiscible fluids A and B with different viscosities. All measures of length are given in chosen units abbreviated by ul as unit length.

4.3.3 Rising gas bubble in a viscous fluid

In the previous section, the applicability of the developed multiphase approach has been tested separately for systems with high density as well as high viscosity differences between the phases. In both cases quantitative agreement with the analytical solutions has been achieved. In the following section, these properties are combined for the simulation of the rising of a compressible gas bubble in a heavier incompressible fluid. Therewith, the accuracy of the developed approach to dynamic multiphase flow with large density and viscosity ratios between the regarded phases is examined.

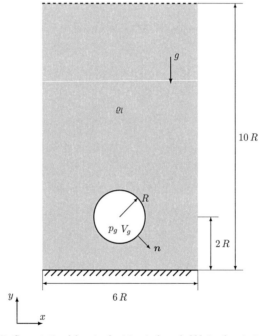

Figure 4.47: Computational domain of a rising single gas bubble in a heavier incompressible liquid matrix.

The computational domain of the test case is sketched in figure 4.47, with an initially circular compressible gas bubble being fully immersed in the incompressible liquid phase.

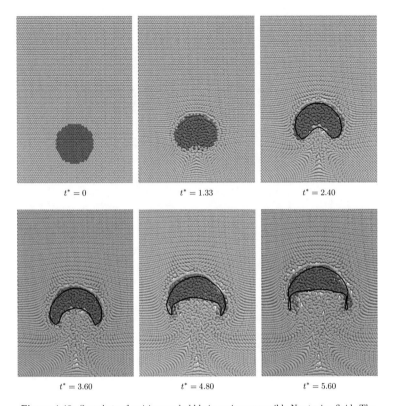

$t^* = 0$ $t^* = 1.33$ $t^* = 2.40$

$t^* = 3.60$ $t^* = 4.80$ $t^* = 5.60$

Figure 4.48: Snapshots of a rising gas bubble in an incompressible Newtonian fluid. The compressible gas bubble is represented by dark gray particles; the incompressible fluid phase by light gray particles. As a reference, the interface position obtained by a grid-based Level-Set approach simulation of two incompressible fluids by Sussman *et al.* is depicted by the black solid line [275].

The dimensions of the domain as well as all parameters are chosen in accordance with the work of Sussman *et al.* for ease of comparison [275]. The gas bubble with its initial radius R starts to rise in the initially static heavier fluid. The domain is ten bubble radii high and six radii wide and bounded by a rigid tank wall at the bottom, while periodic

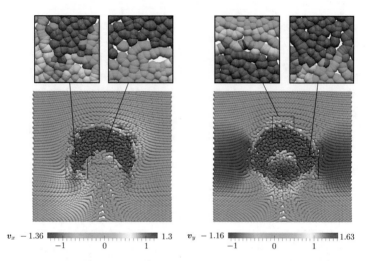

v_x -1.36 ▮▬▬▬▬▬▬▬▬▬▬▮ 1.3 v_y -1.16 ▮▬▬▬▬▬▬▬▬▬▬▮ 1.63
     -1 0 1 -1 0 1

Figure 4.49: Snapshots of the velocity field of the incompressible Newtonian fluid around the rising gas bubble. The compressible gas bubble is depicted by dark gray particles. On the left side the horizontal velocity is shown and the vertical velocity component is sketched on the right. In the details, the sharp interface between the phases is depicted.

boundary conditions are applied in horizontal directions. A free surface condition is applied at the top of the fluid domain. The computational domain is discretized by 6222 particles placed on a hexagonal lattice, with the compressible gaseous phase being represented by 360 particles. The pressure of the gaseous phase is calculated from the gas density as shown in section 3.2 using an equation of state. For the respective application, the relative gas pressure is calculated using Tait's equation of state for sake of comparison with Susman's solution. Tait's equation has been shown in equation (2.84) and is repeated below

$$p_i = \frac{c_g \varrho_{g,i}}{\gamma_g} \left[\left(\frac{\varrho_{g,i}}{\varrho_g^0} \right)^{\gamma_g} - 1 \right], \tag{4.55}$$

with ϱ_g^0 being the reference density of the gaseous particles obtained at the initial configuration. The gaseous phase has a speed of sound of $c_g = 198 \sqrt{g R}$ and a polytropic constant of $\gamma_g = 1.4$ [119]. By using Tait's equation of state, relative pressure values are obtained for the gaseous particles. And since relative pressures are also obtained for the incompressible particles using the projection based approach, no distinction between the phases has to be made in the pressure gradient term. Therewith, the spatial coupling presented in section 3.2.1.1 is obsolete in the current test case and the general corrected gradient discretization as shown in equation (3.56) is applied.

The key parameters in the present test case are the density ratio $\frac{\varrho_l}{\varrho_g}$, which is set to $\frac{\varrho_l}{\varrho_g} = 1000$ and the viscosity ratio $\frac{\eta_l}{\eta_g}$, which corresponds to $\frac{\eta_l}{\eta_g} = 10$ in the present work. Furthermore, the Reynolds number Re has a strong influence on the evolution of the bubble and is defined as [275]

$$Re = \frac{\sqrt{(2\,R)^3 g} \varrho_l}{\eta_l}. \tag{4.56}$$

In the present work, the absolute value of the fluid viscosity is chosen to result in a Reynolds number of $Re = 10$. In addition, the flow field is characterized by the dimensionless Bond number, defined as

$$Bn = \frac{4\varrho_l\,g\,R^2}{\sigma}, \tag{4.57}$$

with σ being the surface tension coefficient. However in the present simulation, surface tension effects are neglected and the Bond number tends towards infinity $Bn \to \infty$. For the presented computation, a cubic spline kernel was used with a smoothing length to particle spacing ratio of $h/L_0 = 1.55$, with L_0 being the initial particle spacing with $L_0 = 0.05$. This results in a cutoff radius of $r_e = 3.1\,L_0$. All computational and material parameters are summarized in table 4.19 below.

In figure 4.48, the rising of the gaseous bubble in the incompressible fluid is shown, starting with the initial condition in the upper left figure. In the second picture at the dimensionless time step $t^* = 1.33$, the rising compressible gas bubble starts to deform and the gas bubble takes an approximately elliptical form. As time proceeds, the characteristic horseshoe shape of the gas bubble develops as it can be seen at the time instant $t^* = 2.40$. For comparison the interface of the same gas–liquid system obtained by a grid based Level-Set approach is superimposed on the particle distribution from now on. The reference interface is indicated by the solid black line and taken form the work of Sussman and coworkers [275]. The evolution of the interface has been obtained by solving the Navier-Stokes equations on an Eulerian grid for two incompressible fluids and the interface position is captured by a Level-Set algorithm. With ongoing time, the gas bubble continues to deform from the horseshoe shape shown at time instant $t^* = 3.6$

to a mushroom-like shape at $t^* = 4.8$ and $t^* = 5.6$. Although surface tension effects are neglected in the SPH approach, while being included in the grid based approach, the agreement between both solutions is satisfactory over the hole time span. Yet, a slight deviation is observed at the forming tails of the gas bubble in the last two pictures. At least a part of the deviation can be ascribed to difficulties in enforcing the periodic boundary conditions for the initial hexagonal particle setup in this particular simulation, which also manifests the slightly non-symmetric evolution of the gas bubble with respect to the center line.

In figure 4.49 the transient velocity field is exemplary depicted at the dimensionless time step $t^* = 3.5$. On the left side, the horizontal component of the velocity vector is shown, while the vertical velocity field, which is responsible for the characteristic deformation of the gas bubble, is sketched on the left. The color coding of the liquid phase corresponds to the respective component of the velocity vector. In contrast, the gaseous particles are colored in gray. Likewise, the local particle distribution at the interface between gaseous and liquid phase is shown in the details at four different positions. At all four positions, the interface between gaseous particles and the liquid particles stays sharp even with the relatively low dynamic viscosities of both fluids and without including surface tension. Though, at the forming tails of the bubble very few particles pierce the interface and move into the opposite phase as it is shown in the respective details. But the stability of the simulation is not affected.

With the presented incompressible – compressible SPH approach, multiphase flow with a large density ratio as well as viscosity ratio between the phases can be simulated in a stable and accurate manner. In addition, the interface stays sharp over the whole simulation time. One has to stress, that no artificial forces as well as additional smoothing steps, like the XSPH approach, are used as it is common in other multiphase SPH approaches [67, 136, 276, 277]. Moreover, no artificial surface forces are applied to keep a sharp interface and maintain a regular particle distribution as it is done in the work of Colagrossi, Das and Das as well as Grenier and coworkers, where the same test case is calculated using a weakly compressible SPH approach [67, 119, 136]. As already stressed several times, the use of any artificial forces is avoided whenever possible in the present work, since any additional surface force or smoothing procedure has to be adjusted very carefully in order to not affect the particle dynamics in excess.

In the present section, the applicability of the developed multiphase approach for buoyancy driven flow has been demonstrated. However, in the desired application, the structure formation is driven by the hydrostatic pressure of the blowing agent. Therefore, the developed liquid–gas interface coupling is validated with the expansion of a single bubble under internal overpressure in the following paragraph.

Parameter		Value
initial particle distance	L_0	$0.05\,ul$
initial bubble radius	r	$10\,L_0$
cutoff radius	r_{cut}	$3.1\,L_0$
surface bound	β^S	0.85
initial fluid density	ϱ_l^0	$1\,\frac{kg}{ul^3}$
initial gas density	ϱ_g^0	$0.001\,\frac{kg}{ul^3}$
gravitational acceleration	g	$9.8\,\frac{ul}{s^2}$
fluid viscosity	η_l	$0.1\,\frac{kg}{ul\,s}$
gas viscosity	η_g	$0.01\,\frac{kg}{ul\,s}$

Table 4.19: Computational details of the rising compressible gas bubble in an heavier incompressible fluid. All measures of length are given in chosen units abbreviated by ul as unit length.

4.3.4 Expansion of a gaseous bubble in a viscous substrate

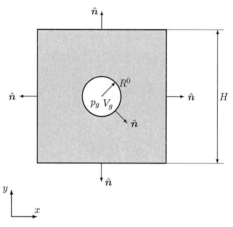

Figure 4.50: Computational domain of the expansion of a single gas bubble in a liquid matrix.

In the previous section, the pressure coupling of the multiphase approach presented in section 3.2 has been validated for a buoyancy driven problem. In the regarded section, the gas – liquid coupling is validated by means of the expansion of a single gas bubble in a liquid Newtonian matrix. This rather simple model allows the derivation of an analytical solution for the expansion of the radius of the gas bubble. Therewith, the verification of the presented SPH approach is possible on a quantitative level. Furthermore, the hybrid SPH approach, especially developed for blowing agent induced structure formation processes and presented in section 3.2.3, will be validated. In the following section, the derivation of the analytical model is sketched, together with the underlying assumptions and simplifications. The computational domain of the test case is depicted in figure 4.50. An ideal gas bubble is embedded in a square shaped, incompressible, liquid Newtonian fluid matrix. The gas and the liquid phase are considered immiscible and surface tension is neglected. Due to the high gas pressure p_g, the gas exerts a stress $-p_g \hat{n}$ on the liquid. Hence, the gas bubble expands and its volume increases, while the gaseous pressure decreases. The steady state is reached when the gaseous pressure p_g is in equilibrium with the external pressure p_{ext}. The simulation is performed in two dimensions and the gaseous bubble is reduced to a disk with an initial radius $R(t) = R^0$, volume V_g^0 and respective gas pressure p_g^0. For the derivation of an analytical solution for the expansion process, the incompressibility of the fluid phase is assumed. In addition, the gas phase is considered as ideal and the viscosity of the gas phase is neglected. For vanishing external pressure, the temporal evolution of the bubble radius can be simplified to

$$R(t) = R^0 \sqrt{1 + \frac{p^0}{\eta} t}, \qquad (4.58)$$

The detailed derivation of the analytical solution is given in appendix H.

In the following, two different approaches for handling the gaseous phase are discussed in the face of the regarded application, the structure formation by release of a blowing agent. In the respective application, the kinematic of the gaseous phase plays a negligible role for the resulting pore structure, since the deformation is driven by the pore pressure. This fact is used for simplifying the overall simulation by neglecting the dynamic of the gas phase and calculating the gaseous pressure and pore volume by means of an underlying grid as described in section 3.2.3. In the second approach, the gaseous phase is resolved in detail within the SPH formulation as presented in section 3.2.1.

In both approaches, the hydrostatic gas phase pressure follows the ideal gas law

$$p_g = p_g^0 \frac{V^0}{V}. \qquad (4.59)$$

with V being the actual volume of particle i and V^0 the initial particle volume. In the hybrid approach V resembles the actual and V^0 the initial pore volume calculated from all connected unoccupied grid cells. The fluid viscosity η_l is equal to unity and the dynamic gas viscosity is $\eta_g = 10^{-4} \frac{kg}{ul\,s}$. The initial gas pressure p_g^0 is set to $p_g^0 = 0.5 \frac{N}{ul^2}$ and the initial radius of the gas bubble is set to $R^0 = 0.1076\,ul$. Further computational details are summarized in table 4.20.

In the following, the simulation results for the expansion of a single gas bubble in the Newtonian liquid are discussed. In order to assess the developed multiphase approaches to pore formation processes driven by a hydrostatic gas pressure, the test case is simulated with the NSPH as well as NCSPH discretization approach. Further, the conventional SPH approach with a simplified liquid–gas coupling is used to show the necessity of the developed improvements. However, before the results of the different methods are compared, the principle dynamic of the expansion is discussed. As initially

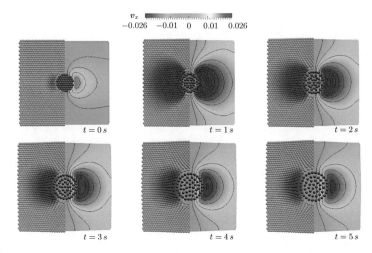

Figure 4.51: Horizontal velocity distribution of the expanding viscous fluid and position of the gaseous particles at different instants in time obtained using the NCSPH approach. The isovalues for the horizontal velocity are shown from $v_x = -0.026 \frac{ul}{s}$ to $v_x = 0.026 \frac{ul}{s}$ in steps of $\Delta v_x = 2 \cdot 10^{-3} \frac{ul}{s}$.

stressed, the single bubble expands over time due to the stress exerted on the fluid by the gaseous phase. The expansion process simulated with the NCSPH approach is exemplary shown in figure 4.51 at six instants in time. In the upper left figure, the initial configuration at time instant $t = 0\,s$ is shown. To ease the identification of the fluid–gas interface, the gaseous particles are sketched in gray in all figures, while the color coding of the liquid particles represents the horizontal velocity. Likewise, the fluid particles are only displayed in the left half of the figures. On the right, the isocontour lines indicate the velocity values starting from $v_x = -0.026 \frac{ul}{s}$ to $v_x = 0.026 \frac{ul}{s}$ in steps of $\Delta v_x = 2 \cdot 10^{-3} \frac{ul}{s}$. As it can be seen in all figures, the respective velocities reach their maximal absolute values at the interface, which is characteristic for an expansion process. In addition, the high quality of the developed gas–liquid interface coupling is demonstrated by maintaining a symmetric and regular particle distribution over the whole simulation time. Due to the low dynamic viscosity of the gas phase, this behavior is especially prominent for the gaseous particles.

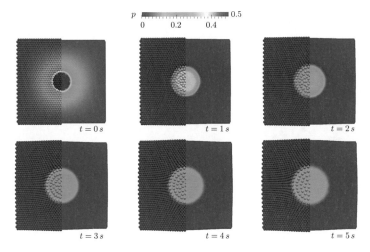

Figure 4.52: Pressure distribution during the expansion of the ideal gas bubble at various time steps obtained using the NCSPH approach.

In figure 4.52 the calculated pressure fields of the liquid–gas system are shown at the same instants in time. As expected, the pressure of the gaseous phase decreases with increasing pore volume. Furthermore, it can be seen, that the Von Neumann boundary condition applied for solving the pressure Poisson equation is satisfied during the whole simulation process. Therewith, the pressure of the liquid interface particles is always slightly higher than the respective values of the neighboring bulk fluid particles. So the corrective velocity calculated from the obtained pressure field does not counteract the expansion velocity of the bubble.

So far, the simulation results have been discussed on a qualitative basis only. Thus, the expansion of the gas bubble is compared to the analytical solution in the following. In figure 4.53 the growth of the dimensionless bubble radius is depicted over time for the different approaches. The results of the NSPH approach are marked by a red triangle (◁) and the one of the NCSPH approach by a blue circle (○). As a reference, the analytical solution is sketched by the solid line. In the left diagram, the growth of the dimensionless radius over time for the purely SPH based multiphase approaches is shown. As expected, almost perfect agreement with the analytical solution is achieved using the Normalized

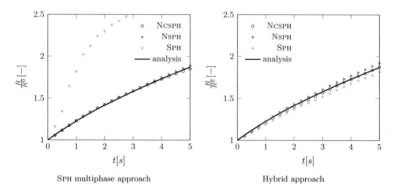

SPH multiphase approach　　　　　　　　　　　Hybrid approach

Figure 4.53: Expansion of the two–dimensional ideal gas bubble of initial radius $R^0 = 0.1076$ in a viscous fluid domain. Comparison of the Ncsph (○), Nsph (◁) and Sph (▽) solution with the analytical solution (solid line).

Corrected as well as Normalized Sph multiphase approach. In contrast, the conventional Sph approach with the conventional gas-liquid interface treatment is not capable of predicting the expansion of the gas bubble as it is depicted in figure 4.53 by the ▽ on the left side. The apparent difference between the conventional Sph method and the two developed approaches can be explained with the help of the snapshots of the particle distribution displayed in figure 4.54. In the figure, the snapshots of the particle distribution during the expansion process are shown for the three different methods, with the color coding representing the normalized density. In the left figure, the particle configuration obtained with the conventional Sph approach at time instant $t = 1\,s$ is depicted. As it can be seen, the fluid density decreases approximately by 40 % towards the liquid–gas interface. This is due to the restricted summation over particles of their own phase for calculating the fluid density as sketched in equation (2.96). In the other two pictures, the density distributions at time step $t = 5\,s$ obtained with the Normalized Sph as well as Normalized Corrected Sph approach are shown. With both approaches, the decrease of the fluid density towards the free surface and the interface is effectively reduced. As expected, the improvement is more pronounced for the Normalized Corrected Sph approach. In addition, it can be seen in the left figure, that the gaseous particles are randomly and irregularly distributed within the gas

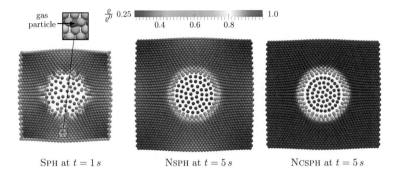

SPH at $t = 1\,s$ NSPH at $t = 5\,s$ NCSPH at $t = 5\,s$

Figure 4.54: Snapshots of the dimensionless density of the gas and fluid phase during the expansion of an ideal gas bubble obtained by the different simulation methods. Left: Particle positions and density contours computed by the conventional SPH approach at $t = 1\,s$. Center: Particle positions and density values obtained by the NSPH multiphase approach at $t = 5\,s$. Right: Positions and density values at $t = 5\,s$ simulated with the NCSPH multiphase approach.

bubble using the conventional SPH approach. This can be explained by the conventional gas-liquid phase coupling. As stated above, the deformation of the compound matrix is mostly governed by the hydrostatic pressure and the gaseous phase is expanding within the evolving pore. Therewith and as given by the kinematic boundary condition, no velocity gradient in normal direction to the interface occurs during the expansion. However, due to the discrete nature of the purely particle based approach and the insufficient simplified gas–liquid phase coupling of the conventional SPH approach, the gas particles are partly moving with excess speed towards the interface, are decelerated and eventually repelled from the interface. One gas phase particle even penetrates the fluid interface in the lower part of the computational domain as it can be seen in the detail of the left snapshot in figure 4.54, while all other gaseous particles are repelled again at the interface. Nevertheless, this acceleration–deceleration process proceeds, introduces an additional stress on the liquid surface and results in the quasi-random and irregular distribution of the gas phase particles. Owing to this behavior, the faster growth of the gas bubble can be explained, since the particles comprising the liquid matrix are also accelerated by the viscous interaction with the gas particles. Thereby, it has been demonstrated that the simplified treatment of the liquid–gas interface, as used in the conventional approach, is not sufficient to model the expansion process. In contrast, the gaseous particles are regularly and evenly distributed in the pore space as sketched in figure 4.54 in the middle and right snapshot using the NSPH as well as NCSPH approach for the simulation of the expansion process. To conclude, both of the developed purely SPH based multiphase approaches using the NSPH as well as NCSPH discretization scheme are capable of describing the expansion of a gaseous bubble in a viscous substrate with high accuracy.

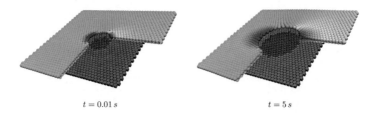

$$t = 0.01\,s \qquad\qquad\qquad t = 5\,s$$

Figure 4.55: Snapshots of the expansion of an ideal gas bubble simulated with the NCSPH hybrid approach. Occupied grid cells are colored in gray, while unoccupied grid cells are marked in red. The color coding of the SPH particles represents the horizontal velocity.

So far, the expansion of the ideal gas bubble in the viscous substrate has been regarded with the multiphase SPH approach. In the following paragraph, the results of the hybrid approach, as described in section 3.2.3, will be discussed at the hand of the present test case. By using the hybrid – SPH approach, the liquid phase is discretized by SPH particles as done before. However, the (simplified) balance equations for the gaseous phase are solved on an underlying grid. As long as the grid cells are occupied by SPH particles, the cells remain inactive. As soon as a grid cell is not blocked by a SPH particle, the grid cell is activated. This is exemplary sketched in figure 4.55, where the particle configuration during the bubble expansion is shown together with the respective grid occupancy. In the first snapshot, the beginning of the expansion process at $t = 0.01\,s$ is depicted. The color coding of the particles represents the horizontal velocity and occupied grid cells are shown in gray. Unoccupied grid cells are colored in red. As one can see, the gas bubble still remains in its original shape and size at that time point. In the second snapshot on the right, the expanded bubble, represented by the expanded unoccupied grid cells, is

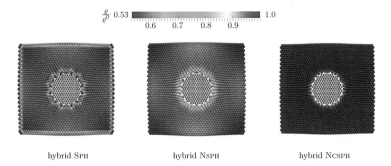

hybrid SPH   hybrid NSPH   hybrid NCSPH

Figure 4.56: Snapshots of the dimensionless density of the fluid phase during the expansion of an ideal gas bubble at $t = 5$. Left: Particle positions and density contours computed by the conventional hybrid – SPH approach. Center: Particle positions and density values obtained by the hybrid NSPH multiphase approach. Right: Positions and density values simulated with the hybrid NCSPH approach.

shown at time $t = 5\,s$. The temporal evolution of the expansion process simulated using the NSPH – and NCSPH – hybrid approach is shown figure 4.53 on the right. With the NCSPH approach, perfect agreement with the analytical solution is obtained, while the NSPH approach overpredicts the expansion slightly. The evolution of the pore radius obtained with the hybrid approach in conjunction with the conventional SPH method is included in figure 4.53. The conventional SPH – hybrid approach is represented by the green nabla (∇) and the results are also in good agreement with the analytical solution. In figure 4.56 the final particle positions and dimensionless density contours are depicted for the three different methods. Using the conventional SPH approach, the density decreases strongly towards the free surface as shown in the left picture. By using the renormalized kernel, the decrease in the density towards the free surface is effectively reduced to approximately $10\,\%$ in the NSPH approach as sketched in the center picture. On the right, the result of the fully corrected NCSPH approach is shown. No significant decrease in the fluid density towards the boundaries is observed in the latter case.

In the present section, the developed purely SPH based multiphase as well as hybrid – SPH approach have been validated for the simulation of pressure driven structure forming processes. In addition, the capabilities of the developed SPH approaches to quantitatively describe complex material have been demonstrated, together with their ability for the simulation of thin free surface dominated structures. Therewith, the foundations for the quantitative simulation of the formation of open-porous materials are laid and the developed approaches are applied to the application example, introduced in section 1.2, in the subsequent section.

Parameter		Value
initial particle distance	L_0	$0.025\,ul$
grid cell length	Δz	$0.025\,ul$
cutoff radius	r_{cut}	$3.1\,L_0$
initial bubble radius	R	$0.1\,ul$
surface boundary	β^S	0.8
initial fluid density	ϱ_l^0	$1\,\frac{kg}{ul^3}$
fluid viscosity	μ_l	$1\,\frac{kg}{ul\,s}$
initial gas density	ϱ_g^0	$1\,\frac{kg}{ul^3}$
gas viscosity	μ_g	$10^{-4}\,\frac{kg}{ul\,s}$
maximum time step	Δt	$10^{-4}\,s$

Table 4.20: Computational details of the simulation of the expansion of a compressible ideal gas bubble in an incompressible Newtonian fluid. All measures of length are given in chosen units abbreviated by *ul* as unit length.

4.3.5 Interim conclusion

In the previous sections, the developed multiphase algorithm for describing the inter-action of compressible as well as incompressible phases has been validated. There it has been shown, that large density as well as viscosity ratios between the phases can be described in a numerically stable manner and with high accuracy. Moreover, the developed approach has been applied to the simulation of the expansion of a single gas bubble in a liquid matrix, which can be seen as a first step towards the simulation of the reaction-induced formation of an open-porous material. Again, the numerical solution is in good agreement with the theoretical prediction. Last but not least, a simplified modeling approach by combining the SPH particle approach with a stationary underlying grid to the so called hybrid approach, as derived in section 3.2.3, has also been applied to the latter test case. Once again, the results are in good concordance with the analytical solution.

5

Simulation of the morphogenesis of open-porous materials

After verifying the validity of the developed SPH approach in section 4 by direct comparison with analytical solutions, experiments and established continuum mechanics models, the developed SPH approach is applied to the pore formation process presented in section 1.2. In the course of this, the following processes have to be considered for modeling the morphogenesis of open-porous materials by release of a blowing agent. First, oxygen diffuses through the compound matrix and, upon reaching the embedded wax islands, the oxygen reacts with wax, decomposes it and forms the gaseous blowing agent. The pressure build-up in the emerging pore by formation of a blowing agent and the resulting deformation of the surrounding heterogeneous material have to be included. A porous network is formed by the successive coalescence of expanding voids. Thus, the evolution of internal and external surfaces has to be considered together with the shortening of the diffusion path through the evolving pores. As a last step to the formation of an open-porous material, the fracture of the polymer matrix has to be described. In the present chapter, the developed approaches are first applied to simulate the reaction-induced formation of a single pore. Subsequently, the emergence of an open-porous transport pore system is modeled. Finally, the influence of various different model parameters on the predicted morphology is discussed in a parametric study.

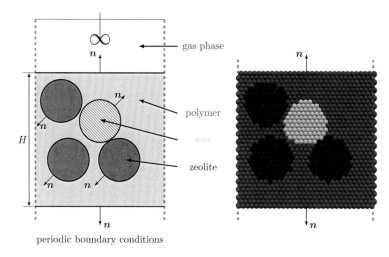

Figure 5.1: Computational domain representing an excerpt of the adsorber monolith (left) and its discretization for the simulation of the reaction-induced formation of a single open pore (right).

5.1 Reaction-induced formation of a single open pore

5.1.1 Normalized Corrected SPH approach

In a first step, the formation of a single pore with the developed multiphase NCSPH approach is presented as a proof of concept. The respective computational domain is shown in figure 5.1. The domain is discretized by 1394 particles, distributed on a hexagonal lattice. The different material types are distinguished by color, with red particles representing the polymer, black particles depicting the zeolite and yellow particles constituting the wax phase. The polymer and wax phase behaves like an incompressible viscoplastic Bingham material as described in section 4.2.2. The zeolite material is modeled as a linear elastic material according to the constitutive equation shown in section 4.2.4. Therewith, the zeolite particles are slightly compressible. The computational details are summarized in table 5.1 below. The purely particle based NCSPH approach as described in section 3.2.1 and 3.2.2.1 is applied. Periodic boundary conditions are set in the horizontal direction. A Dirichlet boundary condition for the

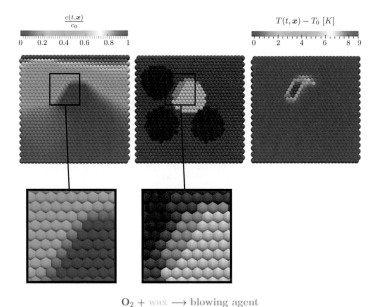

$$O_2 + \text{wax} \longrightarrow \text{blowing agent}$$

Figure 5.2: Modeling of the reaction-diffusion process on the deforming body. Left: Diffusion of oxygen in the heterogeneous compound; Center: Decomposition of wax particles (yellow) with oxygen to form a gaseous blowing agent depicted by green particles; Right: Temperature field.

oxygen concentration is specified at the top free surface. Oxygen is diffusing through the compound from the top and reacts with the wax particles upon contact. In the subsequent section, the reaction-diffusion approach within a deforming body is described in detail.

For modeling the overall process, the decomposition of the wax and the formation of the blowing agent have to be included into the simulation. A simple oxidation reaction is assumed to occur in the wax phase only, with the methene group $(-CH_2-)$ being the representative building block. The mass balance for the wax particles is formulated as follows

$$\frac{D\,m_i}{D\,t} = -k(T)\,c_{O_2,i}\,V_i\,MW_i, \tag{5.1}$$

This process is sketched in figure 5.2, with the dimensionless oxygen concentration shown on the left. As one can see, the oxygen diffuses through the polymer matrix and the zeolite particles. Since a higher diffusion coefficient of oxygen in zeolite is specified in the simulation, the wax phase starts to decompose first on the side turned towards the upper left zeolite particle and the oxygen is completely consumed. In the center of figure 5.2 the particle types are shown, with some wax phase particles already being replaced by green gaseous blowing agent particles. Since the mass of the original wax particle corresponds to the mass of the formed blowing agent, the gas phase pressure can be calculated by an equation of state. For simplicity, the ideal gas law

$$p_i = \frac{\varrho_i \, R \, T_i}{MW_i} \tag{5.2}$$

is used in the present simulation. The gas pressure acts on the surface of the compound, which leads to a deformation of the latter. The evolution of the substrate to an open-porous system is depicted in figure 5.3 from the upper left to the lower right picture. In the first picture, some oxygen already diffused into the polymer matrix, but no wax particles decomposed so far. In the second picture, most of the wax particles are decomposed and the shear rate in the upper region of the polymer matrix exceeds the viscoplastic yield stress due to the stress exerted on the matrix surface. After exceeding the yield stress in a larger domain, the polymer continuous to flow as shown in the third and fours picture. In the fifth picture, the open pore starts to form, while the plastic deformation continues until the pressure compensation with the outside takes place. The final open-porous structure is shown in figure six.

With the developed multiphase NCSPH approach the simulation of the formation of an open-porous material is possible as demonstrated in figure 5.3. In the development of the NCSPH approach, a strong focus was laid on obtaining a consistent numerical scheme, especially with regard to free surfaces but also interfaces, where the interpolation domain can also be truncated. Since the developed NCSPH approach proved very accurate for solving the respective continuum equations and test cases as shown in section 4, the NCSPH multiphase approach is expected to give quantitative results for the evolution of the material structure during the pore formation process. Hence, the NCSPH approach is considered as the benchmark and the simplified NSPH approach, as described in section 3.2.2.2, is compared to the results of the former in the subsequent paragraph.

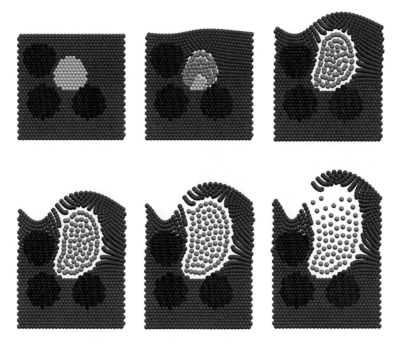

Figure 5.3: Simulation of the morphogenesis from the initial substrate (upper left) to the open-porous material (lower right). Red particles represent the viscoplastic polymer, yellow the viscoplastic oxidizable wax and black particles represent the zeolite material, while the gaseous blowing agent is represented by green particles.

5.1.2 Normalized SPH approach

In the current paragraph, the simulation of the morphogenesis is repeated using the NSPH method with all parameters kept constant. In contrast to the previous results generated with the NCSPH approach, no significant plastic deformation of the compound matrix is observed using the NSPH discretization scheme as it is shown in figure 5.4 on the right. The reason for the deviation lies in the fact, that the viscoplastic yield stress is not exceeded during the simulation. This can be attributed in parts to the treatment of the stress tensor at the free as well as internal surface of the polymer matrix and will be discussed in detail in the following.

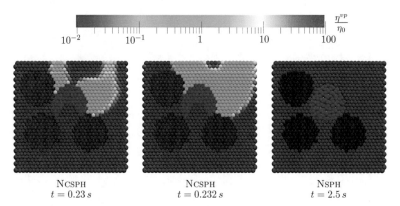

$$\frac{\eta^{vp}}{\eta_0}$$

| 10^{-2} | 10^{-1} | 1 | 10 | 100 |

| NCSPH | NCSPH | NSPH |
| $t = 0.23\,s$ | $t = 0.232\,s$ | $t = 2.5\,s$ |

Figure 5.4: Comparison of the effective viscosities (equation (4.21)) at different time instants calculated by NCSPH (left and center) and NSPH approach (right).

On these grounds, the effective viscosities normalized by the plastic viscosity, as defined in equation (4.21), are depicted for both methods in figure 5.4. In the left and center picture, the effective viscosity obtained by the NCSPH approach at time step $t_1 = 0.23\,s$ and $t_2 = 0.232\,s$ is shown. With increasing time, the effective viscosity of the viscoplastic polymer decreases in the regions above the developed pore due to the stress exerted by the gas phase on the compound matrix. As soon as the viscoplastic yield stress is exceeded locally, the polymer begins to flow in that region. In contrast, using the NSPH approach, the actual stress in the polymer does not exceed the viscoplastic yield stress and as a consequence the effective viscosity does not decrease as shown in the right picture at the time instant $t_3 = 2.5\,s$, even though the same force is acting on the matrix surface. This can be explained by the inconsistency of the NSPH interpolation at the truncated interpolation domain close to the gas–fluid interface. The stress tensor τ_i of the viscoplastic particle i at that interface is calculated as

$$\tau_i^{\alpha\beta} = \frac{1}{\Gamma_i} \left\{ \eta_i^{vp} \sum_{j \in \Omega_i} V_j \left[\left(v_j^\alpha - v_i^\alpha \right) \nabla_i^\beta w_{ij} + \left(v_j^\beta - v_i^\beta \right) \nabla_i^\alpha w_{ij} \right] \right. \tag{5.3}$$

$$\left. + \sum_{j \in \Omega_G} \frac{2\,\eta_i^{vp}\,\eta_j}{\eta_i^{vp} + \eta_j} \left[\left(v_j^\alpha - v_i^\alpha \right) \nabla_i^\beta w_{ij} + \left(v_j^\beta - v_i^\beta \right) \nabla_i^\alpha w_{ij} \right] \right\}, \tag{5.4}$$

with all neighboring particles included regardless of their type. Thereby, the effective viscosity of the viscoplastic material η_i^{vp} of particle i is estimated to

$$\eta_i^{vp} = \eta_i + \tau_o \frac{1 - e^{-m\dot{\gamma}_i}}{\dot{\gamma}_i} \qquad \forall i \in \Omega_F \tag{5.5}$$

And for the calculation of the magnitude of the strain rate tensor $\dot{\gamma}_i$, defined according to equation (4.20) and repeated below

$$\dot{\gamma}_i = \sqrt{\frac{1}{2} \sum_\alpha \sum_\beta \dot{\epsilon}_i^{\alpha\beta} \dot{\epsilon}_i^{\alpha\beta}} \qquad \forall i \in \Omega_F, \tag{5.6}$$

the strain rate tensor $\dot{\underline{\epsilon}}_i$ is predicted for all particle $i \in \Omega_F$ following

$$\dot{\epsilon}_i^{\alpha\beta} = \frac{1}{\Gamma_i^{\Omega_F}} \sum_{j \in \Omega_F} V_j \left[\left(v_j^\alpha - v_i^\alpha \right) \nabla_i^\beta w_{ij} + \left(v_j^\beta - v_i^\beta \right) \nabla_i^\alpha w_{ij} \right], \tag{5.7}$$

with the summation being limited to the neighboring viscoplastic particles only. Since the gaseous particles are not included in the calculation of the strain rate for the effective viscosity, the domain as truncated, which leads to an inconsistent interpolation. Hence, the strain rate tensor $\dot{\underline{\epsilon}}_i$ adopts a lower absolute value, which in turn leads to a higher effective viscosity. And since the effective viscosity changes strongly with variation of the strain rate, as it is exemplary depicted in figure 5.5, the drastic impact on the evolving morphology can be explained. Lastly, it has to be mentioned, that the decrease of the norm of the stress tensor towards the free surface is in accordance with the observation made by Colagrossi *et al.* as well as Grenier *et al.* Both report an unphysical decrease of the stress of a Newtonian fluid and so of the velocity gradient at the free surface for the uncorrected SPH interpolation [119, 218].

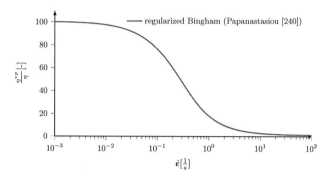

Figure 5.5: Effective viscosity of the regularized Bingham fluid as a function of the strain rate $\dot{\epsilon}$.

Based on these considerations, the ability of the moderately corrected NSPH approach to describe the morphogenesis of open-porous materials on a quantitative basis is questioned. Thus, only the NCSPH discretization approach is used in the following section for the simulation of open-porous materials. However, one has to mention, that the viscoplastic yield stress τ_0 is lying in a region where small differences in the strain rate yield notably different effective viscosities and as a result, notably different structures.

In the present section, both multiphase SPH approaches have been applied for the simulation of the morphogenesis of open-porous materials. In the process, the NCSPH multiphase method proved superior in comparison to the NSPH discretized version. In the subsequent paragraph, the simulation of the morphogenesis of a single open pore is performed using the hybrid NCSPH framework, within which the kinematics of the gas phase is neglected for simplification.

Parameter		Value
initial particle distance	L_0	$0.025\,ul$
cutoff radius	r_{cut}	$3.1\,L_0$
surface boundary	β^S	0.8
maximum allowable time step	Δt	$5 \cdot 10^{-7}$
substrate height	H	$0.72\,ul$
substrate length	L	$1\,ul$
initial polymer density	ϱ_p^0	$1\,\frac{kg}{ul^3}$
initial wax density	ϱ_w^0	$1\,\frac{kg}{ul^3}$
initial zeolite density	ϱ_z^0	$10\,\frac{kg}{ul^3}$
polymer dynamic viscosity	η_0	$1\,\frac{kg}{ul\,s}$
yield stress	τ^{vp}	$13.2\,\frac{kg}{ul\,s^2}$
wax dynamic viscosity	η_w	$1\,\frac{kg}{ul\,s}$
gas dynamic viscosity	η_g	$0.01\,\frac{kg}{ul\,s}$
Hooks modulus zeolite	E	$2.1 \cdot 10^3\,\frac{kg}{ul\,s}$
Poisson's ratio zeolite	ν	0.45
diffusion coefficient polymer	$D_{O_2,p}$	$10^{-2}\,\frac{ul^2}{s}$
diffusion coefficient wax	$D_{O_2,w}$	$10^{-2}\,\frac{ul^2}{s}$
diffusion coefficient gas	$D_{O_2,g}$	$0.1\,\frac{ul^2}{s}$
diffusion coefficient zeolite	$D_{O_2,z}$	$0.1\,\frac{ul^2}{s}$

Table 5.1: Computational details of the simulation of the formation of single pore using the entirely SPH based approach. All measures of length are given in chosen units abbreviated by ul as unit length.

5.1.3 The hybrid – NCSPH approach

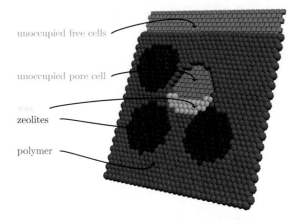

Figure 5.6: Discretization of the computational domain of the formation of a single pore with the hybrid SPH approach. Red particles represent the viscoplastic polymer, yellow the viscoplastic oxidizable wax and black particles represent the zeolite material. Unoccupied grid cells forming a closed pore are marked in green, while the other unoccupied cells are colored in gray.

As stressed in section 3.2.3, the kinematic of the gas phase plays a negligible role for the resulting pore structure. Therefore, the convection of the gas phase is not resolved in the hybrid approach, since the stress exerted on the compound matrix is dominated by the hydrostatic pressure of the gaseous phase. In the following subsection, the formation of a single open pore is calculated with the NCSPH hybrid approach and compared to the results obtained by the purely SPH based NCSPH multiphase approach shown in the paragraph 5.1.1. The discretization of the computational domain in the context of the hybrid approach is shown in figure 5.6. The chemical decomposition is modeled in the same way as in the purely SPH based approach. However, as soon as the mass of a wax particle reaches zero, the particle is completely removed from the calculation. At the same time, the underlying grid cell is activated and the resulting pore volume is calculated from the unoccupied underlying grid cells by counting the number of connected unoccupied cells. These cells are marked in green in figure 5.6. Since the

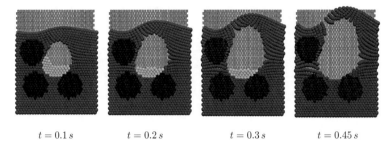

$t = 0.1\,s$ $t = 0.2\,s$ $t = 0.3\,s$ $t = 0.45\,s$

Figure 5.7: Simulation of the morphogenesis from the initial substrate compound as shown in figure 5.1 to the open-porous material using the hybrid approach at different time steps.

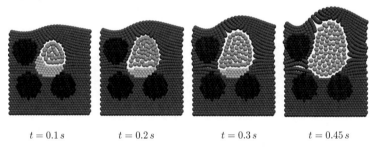

$t = 0.1\,s$ $t = 0.2\,s$ $t = 0.3\,s$ $t = 0.45\,s$

Figure 5.8: Comparative simulation of the previous morphogenesis process with the purely SPH based algorithm under otherwise identical parameters.

initial mass of the wax particle corresponds to the mass of the formed blowing agent, the pressure inside the pore can be obtained by an equation of state since the pore volume is known. The molecular transport in the evolving pore system is also modeled on the underlying grid.

In the following, the simulation of the morphogenesis towards an open-porous material using the hybrid NCSPH approach is discussed. The same initial configuration as in the previous case and shown in figure 5.1 is used. Though, the diffusion coefficients in the compound are altered in comparison to the previous simulation. The respective computational details are summarized in table 5.2. In figure 5.7, the temporal evolution of the compound towards an open-porous material is shown. In all pictures, grid cells composing the closed pore filled with the blowing agent are marked in green, while the rest of the unoccupied grid cells are tagged in gray. In the first picture, the particle distribution after 0.1 time units is shown together with the underlying grid. Approximately two third of the wax island is already oxidized by the permeating oxygen and the gaseous blowing agent is already formed. A beginning plastic deformation is observed in the upper region of the compound material. In the second picture, the pore begins to widen considerably. Therewith, more and more underlying grid cells are left unoccupied and are activated for the interaction with the SPH particles. With continuing decomposition of the wax phase, the stress in the polymer material increases, leading to an increased plastic material deformation as shown in the third picture. With increasing time, the closed pore is widened, as depicted by the green cells, until the open pore is formed. As soon as a connection with the outside cells in established, the pressure equalizes, and the final structure of the compound material is shown in the last figure. Since an open pore is established, the underlying unoccupied grid cells comprising the pore are marked in gray.

To assess the results of the simplified hybrid approach, the simulation is repeated using the entirely SPH based multiphase NCSPH algorithm with otherwise identical parameter. The results of the comparative simulation are shown in figure 5.8. Thereby, the particle configuration at the corresponding time steps as illustrated above in figure 5.7 are depicted. Almost perfect agreement between both simulation approaches can be observed for the first three time instants ($t = 0.1\,s, 0.2\,s, 0.3\,s$). Yet, a clear discrepancy is observed between both last picture of each series. In case of the hybrid approach, the open pore has been already formed after $t = 0.45\,s$ as shown in the last snapshot. And this is not the case using the purely particle based approach, where the overall particle configuration and viscoplastic deformation is comparable to the one obtained by the hybrid approach, but the pore has not opened yet within that time frame.

That deviation is obviously attributed to the different treatment of the gaseous phase. As described in section 3.2.3, the interaction between the gaseous phase and the compound matrix is modeled by a body force vector acting on the compound surface. As shown in equation (3.113), the body force per unit mass is calculated by

$$\boldsymbol{f}_{i \in \Omega_p} = -\frac{p_g \boldsymbol{n}_i}{\rho_i}, \tag{5.8}$$

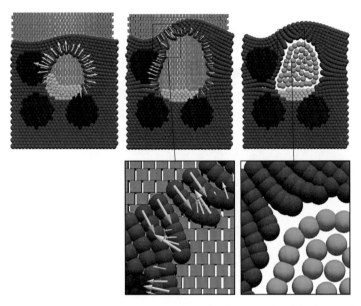

Figure 5.9: Evolution of the force vectors f_i calculated as part of the hybrid approach during the pore formation process shown at the two instants in time. On the left, the particle configuration and the respective force vectors are shown at $t = 0.1$ time units and at $t = 0.3$ time units in the center picture. Below, the detail of the ligament with the force vectors acting on the compound surface at the latter time instant is shown. On the right, the corresponding purely particle based configuration at the same instant in time is shown. In the detail below, the respective particle configuration at the interface is shown.

with the pressure of the gas phase p_g being computed from the number of moles of the blowing agent enclosed in the pore and the volume of the respective pore. And the amount of surface area is included in the expression by the surface normal vector term n_i of particle i, evaluated by equation (3.114) as

$$n_i = \sum_{j \in \Omega_p} V_j \nabla_i w_{ij}. \tag{5.9}$$

In this way, as soon as a SPH particle, representing the compound, is interacting with an activated underlying grid cell, the respective force vector F_i is affecting that particle.

This approach works well during the initial state of pore forming and growing process as it can be seen in the left picture of figure 5.9. The displayed force vectors are all pointing away from the pore space and are acting towards the compound surface. However, two issues arise with ongoing expansion of the pore, which are both based on the current calculation of the surface normal vector. The first issue, which is considered as the main cause for the deviation between both methods, lies in the different working point of the force evoked by the gas phase and acting on the compound surface. That difference can be clearly seen by comparing both details in figure 5.9. While the local surface configuration is taken into account during the calculation of the force vectors within the hybrid approach, the gaseous particles, driving the expansion in the purely particle based method, are located in a certain distance to the compound surface. Therewith, the emerging fracture is not resolved by the gaseous particles. In contrast, the local formation of the crack is resolved within the hybrid approach and so the further opening of the pore is quickened due the increased velocity differences in horizontal direction. The second issue occurring with ongoing expansion of the pore lies in the turn of the orientation of the surface vector for particles located at the outer free surface as it is observed in figure 5.9 in the center and in the extract of the thin ligament below. Since, the ligament between the pore and the upper free surface is thinned with increasing time, particles positioned at the free surface begin to interact with the activated underlying grid cells. Hence, the absolute value of the surface normal vector of these particles as well as its orientation is mostly governed by the free surface. For the particles located directly at the free surface, the normal vector n_i and in this way the force vector f_i is pointing in the opposite direction and so, contrary to the actual expansion trend. Nevertheless, the expansion of the pore is still driven by the particles in the vicinity of the pore, but the opening of the pore is slowed down due to the artificial counter force. Based on that, the current calculation of the surface normal vector based on the summation over the SPH particles as shown above works well for the description of the formation of closed pores as in the initial stages of the latter example and in the test case presented in section 4.3.4. Thus, the approach is not suited for the formation of open-porous materials. For this purpose, a modified approach for calculating the surface normal vector n_i based on the unoccupied grid cells comprising the regarded pore is suggested as

$$\boldsymbol{n}_i = \sum_{k \in \Omega_g} V_k \nabla_i w_{ik}. \tag{5.10}$$

And since only the unoccupied grid cells k, which are part of the respective pore Ω_g, as shown in figure 3.10, are included in the summation, the influence of the free surface on calculating the normal vector is eliminated. Under the presumption of a regular and symmetric particle and grid configuration around the interface, both approaches result in normal vectors with the same absolute value but contrary orientation. This is due to the fact, that

$$\sum_{j \in \Omega_p} V_j \nabla_i w_{ij} + \sum_{k \in \Omega_g} V_k \nabla_i w_{ik} = 0 \tag{5.11}$$

for a particle i located at the compound surface. For sake of completeness, one has to mention, that the above simulation has been repeated with the modified surface tension vector. And as expected, the time until the opening of the pore is slightly reduced, while the overall particle configuration is only slightly changed and therefore not shown. To conclude, the applicability of the developed hybrid approach to the simulation of the reaction induced formation of an open pore has been attested and compared to the results obtained by the purely particle based NCSPH approach. Very good agreement between the results of both approaches has been found starting from the initial formation phase of the pore and the subsequent widening phase. Yet, a rather significant deviation has been observed in the finial stage, the formation of the crack in the ligament between the pore and the outside. The deviation has been attributed to the different exploitation of the evolving compound surface and the emerging crack, which is not geometrically resolved by the SPH gas particles in the purely particle based approach, but taken into account in the hybrid SPH approach. In order to get congruent results with both methods, two enhancements are proposed. On the side of the purely particle based approach, the influence of an increased resolution of the interface by a local adaption of SPH particles, especially for the gas phase, should be investigated. By splitting one particle e.g. into two or four small particles, the resolution of the SPH scheme in general and the resolution of the expanding interface in particular could be improved. As a dynamic adaption criteria, the particle number density should be used. As soon as the number of neighbors of the regarded particle drops below a certain threshold, the splitting will be triggered. Though, it has been shown by Feldman and Bonet, that the dynamic adaption is not trivial, since the superposition of the kernel functions of the newly generated particles at their positions needs to resemble the original kernel function in order to maintain a stable numerical scheme [278]. On the side of the hybrid approach, the impact of the local particle configuration on the calculation of the surface normal vector and thus on the force exerted on the surface, could be reduced by an additional smoothing step over all normal vectors of the neighboring particles $\boldsymbol{n}_j$ as shown below

$$\boldsymbol{n}_i^* = \sum_{j\in\Omega_S\cup\Omega_F} V_j\boldsymbol{n}_j\widetilde{\nabla}_i\widehat{w}_{ij}^{\Omega_F\Omega_S}. \tag{5.12}$$

This additional smoothing step is proposed, since the same procedure is also used for the calculation of the surface tension force between liquid phases based on the CSF approach [279–281]. Here it has been also observed, that the influence of the local configuration on the resulting interface force is overpredicted under certain circumstances. Based on these considerations, both approaches are equally suited for the simulation of the reaction-induced generation of open-porous materials by release of a blowing agent. However, a more detailed description of the evolving fracture of the ductile polymer has to be included in the SPH based framework in a subsequent work. Nevertheless, it has been found in the present section, that both fully corrected NCSPH approaches yield comparable results and are capable of predicting the morphogenesis of open-porous materials. Hence, the approach is extended to simulate the generation of a pore system, where the coalescence of voids is included in the following paragraph.

Parameter		Value
initial particle distance	L_0	$0.025\,ul$
cutoff radius	r_{cut}	$3.1\,L_0$
surface boundary	β^S	0.8
maximum allowable time step	Δt	$5 \cdot 10^{-7}$
substrate height	H	$0.72\,ul$
substrate length	L	$1\,ul$
initial polymer density	ϱ_p^0	$1\,\frac{kg}{ul^3}$
initial wax density	ϱ_w^0	$1\,\frac{kg}{ul^3}$
initial zeolite density	ϱ_z^0	$10\,\frac{kg}{ul^3}$
polymer dynamic viscosity	η_0	$1\,\frac{kg}{ul\,s}$
yield stress	τ^{vp}	$13.2\,\frac{kg}{ul\,s^2}$
wax dynamic viscosity	η_w	$1\,\frac{kg}{ul\,s}$
gas dynamic viscosity	η_g	$0.01\,\frac{kg}{ul\,s}$
Hooks modulus zeolite	E	$2.1 \cdot 10^3\,\frac{kg}{ul\,s}$
Poisson's ratio zeolite	ν	0.45
diffusion coefficient polymer	$D_{O_2,p}$	$10^{-2}\,\frac{ul^2}{s}$
diffusion coefficient wax	$D_{O_2,w}$	$10^{-2}\,\frac{ul^2}{s}$
diffusion coefficient gas	$D_{O_2,g}$	$1\,\frac{ul^2}{s}$
diffusion coefficient zeolite	$D_{O_2,z}$	$0.1\,\frac{ul^2}{s}$

Table 5.2: Computational details of the simulation of the formation of single pore using the hybrid mesh – SPH approach. All measures of length are given in chosen units abbreviated by ul as unit length.

5.1.4 Interim conclusion

In the previous section, the simulation of the formation of a single open pore by release of a blowing agent has been performed with the proposed SPH algorithms. It has been shown, that the viscoplastic nature of the deforming polymer backbone of the adsorber compound puts high requirements on the consistency of the discretization scheme at internal as well as external surfaces. Based on that, the ability of the simplified NSPH approach to describe the evolution towards an open-porous system on a quantitative basis is questioned in its current version. In contrast, quantitative results are expected by the simulation of the overall process by the fully corrected NCSPH approach, which has shown a very high accuracy in the verification of all underlying single processes in section 4. This expectation has further been confirmed, since both NCSPH discretized approaches, namely the purely particle based NCSPH approach as well as the hybrid – NCSPH approach, lead to a very similar evolution of the material structure over time. However, a deviation between both final structures has been observed. That deviation has been attributed to the last step towards the formation of an open-porous structure, the fracture of the thin ligament between the pore and the outside. It has been shown, that the final fracture step is modeled differently in both approaches and two suggestions for the enhancement of both approaches in order to achieve congruent results in the future have been made.

Based on the presented results, no clear recommendation can be given whether the hybrid – NCSPH or purely particle based NCSPH multiphase algorithm should be used. In principle, both methods are equally suited for the simulation of the morphogenesis of open-porous materials. Though, the hybrid – NCSPH algorithm is only partly applicable for the simulation of the formation of an open-porous network with large pore volumes and a broad pore volume distribution. This is due to the current algorithm for detecting the evolving pores. For large pore networks, the calculation of the volumes of the evolving pores starts to be very demanding in a computational costs point of view. For that reason, the generation of an open-porous transport pore system will be conducted with the purely particle based NCSPH approach in the subsequent section.

5.2   Generation of an open-porous transport pore system

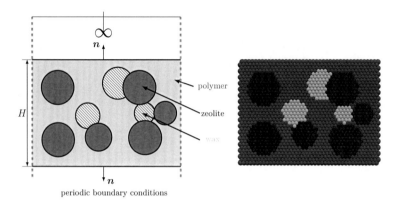

Figure 5.10: Left: Computational domain of the reaction-induced formation of an open-porous transport pore system. Right: Initial configuration and color coding of the particle types.

In the present section, the simulation of an open-porous transport pore system will be conducted. Therewith, not only the formation of an open pore, but also the coalescence of evolving pores has to be considered. In figure 5.10 the computational domain is shown together with the initial particle distribution. The color coding of the particle matches the one in the previous examples. Also similar to the example shown in the paragraph above, a Dirichlet boundary condition for the oxygen concentration is specified at the upper boundary of the compound. Furthermore, periodic boundary conditions are applied in horizontal directions. Further computational details are given in table 5.3. The results of the simulation are depicted in figure 5.11. In the first picture, oxygen already diffused into the compound matrix and part of the wax in the first island is decomposed to the gaseous blowing agent. In the second picture, the polymer is plastically deformed and the wax in the other two islands in the center of the substrate also begins to decompose. In the third picture, the open pore is formed. With it, the diffusion path of the oxygen is shortened. In the fourth picture, the second pore begins to grow, while still no coalescence occurs. No plastic deformation is observed in the vicinity of the third pore on the right side. In the fifth picture, coalescence between

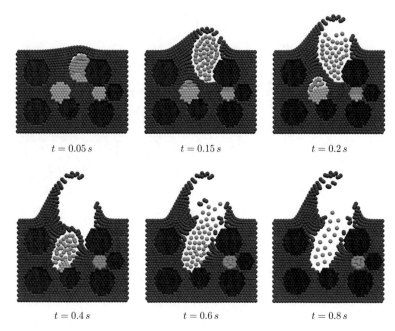

$t = 0.05\,s$ $\qquad\qquad$ $t = 0.15\,s$ $\qquad\qquad$ $t = 0.2\,s$

$t = 0.4\,s$ $\qquad\qquad$ $t = 0.6\,s$ $\qquad\qquad$ $t = 0.8\,s$

Figure 5.11: Simulation of the morphogenesis from the initial substrate compound (upper left) to the open-porous system (lower right).

the two pores has taken place and an open-porous transport pore system towards the zeolites is established. In the last picture, the final open-porous geometry is shown. Still, no considerable deformation is observed for the small pore on the right. Due to the low amount of wax and therefore low build-up pressure in the pore, the exerted stress is not high enough to overcome the viscoplastic yield stress.

To conclude, all characteristics of the pore formation process, as described in section 1.2, have been reproduced in the example shown above. First, oxygen diffuses through the compound matrix and forms a gaseous blowing agent by chemical decomposition of the low molecular wax phase at distributed locations in the heterogeneous compound. Partially large material deformations are caused by the blowing agent, leading to coalescence of pores and the formation of an open-porous structure. In addition, the diffusion path of oxygen is shortened by the shaping of the porous system. Moreover, it has been shown, that an optimal distribution of wax islands can be determined with the developed method. In the following section, a parametric study regarding the influence of the polymer and wax phase properties on the resulting structure is conducted and the influence of the diffusion rate on the structure formation is sketched.

Parameter		Value
initial particle distance	L_0	$0.025\,ul$
cutoff radius	r_{cut}	$3.1\,L_0$
surface boundary	β^S	0.8
maximum allowable time step	Δt	$5 \cdot 10^{-7}$
substrate height	H	$0.72\,ul$
substrate length	L	$1\,ul$
initial polymer density	ϱ_p^0	$1\,\frac{kg}{ul^3}$
initial wax density	ϱ_w^0	$1\,\frac{kg}{ul^3}$
initial zeolite density	ϱ_z^0	$10\,\frac{kg}{ul^3}$
polymer dynamic viscosity	η_0	$0.5\,\frac{kg}{ul\,s}$
yield stress	τ^{vp}	$8.0\,\frac{kg}{ul\,s^2}$
wax dynamic viscosity	η_w	$0.5\,\frac{kg}{ul\,s}$
gas dynamic viscosity	η_g	$0.01\,\frac{kg}{ul\,s}$
Hooks modulus zeolite	E	$2.1 \cdot 10^3\,\frac{kg}{ul\,s}$
Poisson's ratio zeolite	ν	0.45
diffusion coefficient polymer	$D_{O_2,p}$	$10^{-2}\,\frac{ul^2}{s}$
diffusion coefficient wax	$D_{O_2,w}$	$10^{-2}\,\frac{ul^2}{s}$
diffusion coefficient gas	$D_{O_2,g}$	$1\,\frac{ul^2}{s}$
diffusion coefficient zeolite	$D_{O_2,z}$	$0.1\,\frac{ul^2}{s}$

Table 5.3: Computational details of the simulation of the formation of an open-porous transport system using the purely particle based NCSPH multiphase approach. All measures of length are given in chosen units abbreviated by ul as unit length.

5.3 Influence of the polymer, solid and wax phase properties on the structure formation

In the previous section, the applicability of the developed NCSPH approach for the simulation of the morphogenesis of an open-porous material was shown. Based on the developed approach, the influence of the variation of different material properties of the components is exemplary investigated in a parametric study. Thus, the principle ability of the particle based method for describing the manufacturing of open-porous materials under different process conditions is demonstrated in the following paragraph.

In figure 5.12 the resulting final morphologies of the adsorber compound under varied material parameters are depicted together with the initial particle configuration. In order to illustrate the impact of the variation of each material property, the final open-porous material structure from the process presented in the above section 5.2 is depicted as picture C (center, lower row). Starting from the initial particle configuration in the upper center picture, the impact of an increased viscoplastic yield stress τ_0 for the polymer phase is shown on the left picture A. By using a different polymeric material with a yield stress increased by a factor of 10 in comparison to the original polymer (picture C), no open-porous material is formed. Due to the high yield stress, the stress exerted on the polymer surface by the blowing agent is not sufficient to trigger the viscous deformation of the polymer. In contrast, by using a polymer backbone material with a far lower yield stress value ($\tau_0 = 0.1\,\tau_0^{ref}$), viscous flow of the polymer is experienced throughout the substrate. This results in very large material deformations. Due to the reduced yield stress, even the small pore on the right side of the compound is expanded by the blowing agent, which is not the case in the reference simulation.

Besides the polymer characteristics, the resulting material structure is also influenced by the properties of the solid zeolite phase. Exemplary, the density of the solid ϱ_S is reduced by a factor of 10 in comparison to the reference simulation. The resulting morphology is shown in picture D, right next to the reference structure in figure 5.12. With the decreased solid density, which is now of the same order as the polymer density, the inertia of the zeolite particles is reduced and a more uniform stress field is obtained. Therewith, the polymer yield stress is exceeded on a larger domain. This leads in turn to more pronounced material deformations, as can be seen by the larger pore volumes of the two coalescent pores in the center of the compound. Furthermore, the closed pore on the right of the compound is also expanded in comparison to the reference simulation.

Last but not least, the molecular wax is interchanged with a thermoplastic polymer with a higher molecular weight. Thereby, less moles of gaseous blowing agent are formed during the decomposition and the resulting pressure build-up in the pores is decreased. The resulting pore structure is shown in picture E of figure 5.12. The pressure build-up in the surface near pore is sufficient to form an open pore. In contrast, the second pore in the bulk of the compound stays closed, since the reduced pressure is not sufficient to exceed the viscoplastic yield stress. For this reason, the subsurface pore stays closed.

Even though the modified process conditions in the presented examples have been chosen for illustrative purposes, the possibility of the developed particle based simulation approach for predicting the resulting material structure for different material properties as well as manufacturing conditions is demonstrated. Consequently, the developed particle based approach can be applied for optimizing process conditions and support in the selection of the most suitable materials. Thus, a simulation based method capable to improve the development of tailor-made open-porous functional materials has been educed.

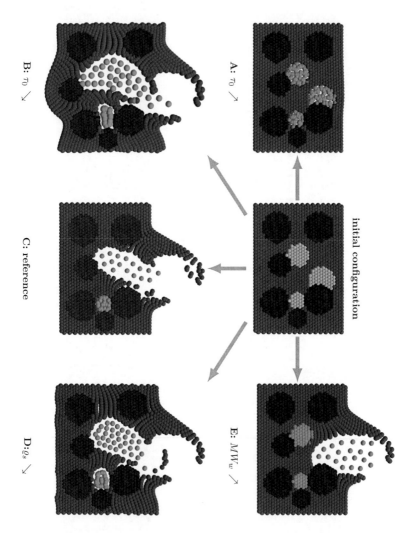

Figure 5.12: Parametric study of the influence of the variation of different material properties on the resulting material structure.

6

Conclusion and outlook

6.1 Conclusion

Open-porous functional materials are utilized in various applications within the field of chemical engineering. Though, the development of the manufacturing process for these functional materials is often very complex and therefore, mostly driven by empirical correlations and experimental experience. This is precisely why, a model based assistance for the development process is highly desired. In the work at hand, a meshfree simulation framework based on the Smoothed Particle Hydrodynamics method has been developed particularly for studying reaction-induced structure formation during the manufacturing of open-porous materials by release of a blowing agent. Due to the Lagrangian nature of the particle based method and the design of the developed approach, the present method is able to accurately describe the reaction–diffusion process on an evolving heterogeneous multiphase body with complex material behavior and large material deformations, fragmentation of ligaments, coalescence of voids as well as interaction of compressible and incompressible phases and between phases with different material behavior.

During the course of this work, two algorithms for describing structure formation processes based on the SPH method have been developed, which mainly differ by the degree of consistency of the SPH approximation scheme. The main characteristics of these methods are summarized in the following. Both approaches, namely the NCSPH and NSPH approach, are based on the conservation equations of mass, momentum and energy. The governing equations are formulated in a Lagrangian way and are discretized based on the concept of the Smoothed Particle Hydrodynamics method. Therewith, the motion of the particles is derived from the laws of continuum mechanics and, in contrast to several other particle models, e.g. the Discrete Element Method, no additional parameters are introduced for describing the interaction between the particles.

In both approaches, a strong focus is laid on the accurate discretization of free surfaces and interfaces as well as the evolution of interacting phases. For the discretization of the multiphase system, different types of particles are used. Each particle type can differ in the utilized constitutive relation, whereof viscous, viscoplastic and viscoelastic as well as linear elastic material models have been implemented and verified. In addition, the flow of the liquid phase particles is modeled as incompressible using a projection based approach, while the compressibility is taken into account for the gaseous as well as solid phase particles. Due to the Lagrangian character of the SPH method, the kinematic interface condition is intrinsically satisfied. However, a special treatment for enforcing the dynamic interface condition between the phases has been developed due to the mixed interaction of absolute pressure values, obtained from the ideal gas law, and relative pressure values, e.g. estimated by the pressure Poisson equation within the projection scheme. Besides these common grounds, both approaches differ considerably in the applied discretization schemes for the governing equations. This is due to the fact, that the conventional Smoothed Particle Hydrodynamics kernel interpolation is not consistent if the support domain is truncated by boundaries, like it is the case at free surfaces and partly at interfaces. Yet, for the simulation of open-porous materials with the appearance of thin structures and ligaments as well as coalescence of voids, these truncated interpolation domains can be predominating. In the NCSPH approach the consistency of the kernel interpolation at truncated domains is restored for constant functions by using a renormalized kernel function. Besides, by using a fully corrected first order derivative the gradient of any linear vector field is correctly evaluated. Due to the applied corrections, the boundary conditions of the balance equations can be directly stipulated on the respective boundary particles within the NCSPH framework. Since the applied corrections increase the computational demand of the SPH algorithm, the simplified moderately corrected NSPH approach has been developed in order to evaluate the necessity of the different corrections for the application at hand. Similar to the NCSPH approach, a renormalized kernel function is used within the kernel interpolation.

Yet, the first order derivative is only partly corrected by the constant denominator of the Shepard kernel. In this way, the computational demand of the SPH scheme is reduced with the drawback, that the first order consistency of the gradient scheme is not ensured at truncated interpolation domains. Additionally, a simplified virtual particle approach is used within the NSPH framework at the free surface to enforce the mathematical boundary condition, especially for the evaluation of the Poisson equation. Both methods have in common, that no artificial forces between particles of their own phase or across the phases are needed to guarantee stability as it is common within the WCSPH methods[1]. As well, no XSPH velocity correction needs to be used in both approaches to prevent inter-penetration of particles.

In order to reduce complexity as well as numerical expense especially for the simulation of the formation of open-porous materials by release of a blowing agent, a so called hybrid approach has been developed in addition. Since the deformation of the compound matrix is mostly governed by the gaseous static pressure in the pores, the actual kinematic of the gaseous blowing agent is neglected and only the compound matrix is discretized using SPH particles. The temporal evolution of the pore volume and gas pressure is calculated on a static underlying grid.

The applicability of the developed SPH approaches for the simulation of open-porous materials has been verified at the hand of several test cases. The focus has been laid on test cases, which verify, respectively validate at least one of the specific underlying single processes occurring during the reaction-induced structure formation by release of a blowing agent or one of its characteristics. At first, the developed basic algorithm is verified with an emphasis on the reconstruction of the interpolation domain at free surfaces. The simplified virtual particle approach proved effective in reducing the errors originating from the truncated domain. However, the most accurate numerical results as well as most stable numerical scheme has been obtained using the fully corrected and normalized NCSPH approach. This is especially important for the description of thin structures with predominating boundary effects, as they occur in the application example at hand. For these kind of simulations, the NSPH approaches proved to be unsuited. Moreover, the applicability of both approaches for describing viscous, viscoelastic as well as viscoplastic and elastic material behavior has been verified. Again,

---

[1]An exception is the collision of the elastic rubber rings test case, where a conservative smoothing approach is applied to suppress numerical instabilities.

251

the NSPH approach was outperformed by the NCSPH approach with varying degrees. Furthermore, the developed coupling between gaseous and liquid phases used in both purely particle based approaches has been also successfully verified by comparison with analytical expressions as well as literature data. This is also true for the developed hybrid approach, where a similar order of accuracy has been obtained.

After the validity of the developed approaches has been verified by direct comparison with analytical solutions, experiments and established continuum mechanics models, the NSPH and NCSPH approaches have been applied to the reaction-induced pore formation process by release of a blowing agent. At first, the formation of a single open pore has been investigated. The regarded application example exhibits all characteristics of a pore formation process, including the diffusion of the educt through the heterogeneous compound matrix and the local generation of the gaseous blowing agent by chemical decomposition of the wax phase. Therewith, partially large material deformations of the viscoplastic matrix are caused, which lead ultimately to the formation of an open-porous structure. The complete process has been simulated using the NCSPH approach. Due to the verification of the algorithm and with the obtained accurate results, the NCSPH based approach has been considered as the benchmark. Based on that, it has been shown, that the NSPH approach is not capable of simulating the investigated structure formation process. The arising discrepancy to the NCSPH solution is attributed to the inconsistent treatment of internal and external interfaces within the NSPH framework. Last but not least, the hybrid – NCSPH approach has been applied to the application case and the result has been compared to the one obtained with the purely particle based NCSPH method. Both simulation results show very good agreement, with an increasing deviation with the start of the crack formation towards the end of the process. This discrepancy has been attributed to the more localized point of action of the force vector in the hybrid – NCSPH approach. Based on the analysis, future optimization possibilities have been derived. Moreover, the simulation of an open-porous transport pore system has been conducted where the coalescence of evolving pores and the shortening of the diffusion path had to be considered. Finally, a parametric study regarding the influence of the polymer, zeolite and wax phase properties on the resulting structure has been conducted and the impact of the diffusion rate and local distribution of the wax islands on the structure formation has been sketched. Thereby, the applicability of the developed SPH algorithms to describe the formation of open-porous materials on a quantitative level has been demonstrated. and the developed particle based approach can be applied for optimizing process conditions and facilitate the selection of the most suitable materials. Thereby, a simulation based method capable to support the development of tailor-made open-porous functional materials has been educed in the work at hand.

6.2 Outlook

Even though the present numerical schemes are able to describe multiphase and multi-physics systems with free surfaces in an accurate manner, improvements can be made in a numerical as well as physical modeling point of view. In the last paragraph further suitable application cases are sketched.

6.2.1 Possible future developments

Extension to three-dimensions

The developed SPH approach has been implemented for two-dimensional systems. For the predictive simulation of the morphogenesis of open-porous materials, a dynamic three-dimensional simulation of all relevant processes is inevitable. Hence, the developed approaches need to be generalized to deal with three-dimensional systems in the next step.

Efficiency of the Corrected Normalized SPH approach

With extending the present NCSPH approach to three-dimensions, the efficiency of the overall algorithm in general and the correction steps of the kernel derivative in particular needs to be improved. As a first step, the correction of the derivatives should be limited to particles whose interpolation domain is truncated by the boundaries only. The implementation is straight forward, since the correction only needs to be triggered for particles having boundary particles in their interpolation domain, respectively neighborhood list.

Extension to large particle numbers and parallelization

By extending the computational domain to three-dimensions and to larger sizes in general, the present code needs to be parallelized. In a first step, a hybrid MPI–OpenMP approach should be taken. As a next step, the performance could be further enhanced with the implementation of a one directional domain decomposition. The latter can be extended stepwise to all three spatial dimensions.

Implementation of a dynamic refinement by particle splitting and variable smoothing length

With increasing pore size, the pore space might not be sufficiently resolved using the current purely particle based SPH approach. In order to increase the resolution in the pore space and therewith allow a more localized pressure distribution, the current SPH framework should be enhanced by including a dynamic refinement of the particle resolution by particle splitting as described in section 5.1.3 [278]. Additionally, a variable, particle specific smoothing length needs to be implemented within the framework. For the particle interactions, the arithmetic mean of the value of the kernel function of both interacting particles has to be used to guarantee the symmetry in the momentum and energy equation [49, 84, 89]. If a more accurate interpolation between the particles of different size is needed, the governing equations need to be developed further by inserting the so called ∇h terms [50, 282, 283]. However, the numerical expense of the overall scheme is considerably increased [89].

Experimental validation

In a next step, simple and well defined experiments of the formation of open-porous materials by release of a blowing agent should be performed in order to validate the developed SPH approach.

Additional material and fracture models

Based on the performed validation experiments, additional constitutive relations need to be implemented within the developed SPH framework. In addition, the current treatment of material fracture is purely mathematically based. As soon as the distance between two particles exceeds the cutoff radius of the kernel function, the fracture has occurred. It has to be evaluated if a more physically sound treatment of the fracture process of the ductile polymer is needed.

6.2.2 Additional applications

Simulation of reacting bubble swarms in bubble column reactors

The multiphase flow in industrial bubble column reactors is inherently complex and various underlying physical and chemical processes have to be considered. Furthermore, typical bubble columns contain many millions of bubbles, which cannot be resolved at acceptable computational costs. Nevertheless, the developed multiphase SPH approach would offer a certain advantage over the grid-based methods for modeling the respective flow. In contrast to the latter, no interface resolving methods are needed due to the Lagrangian nature of the SPH method as stated in section 1.3. However, the particle resolution cannot be reduced to resolve the smallest bubble sizes. For that reason, the NCSPH based multiphase approach needs to be refined by incorporation of a Two-Fluid model, which is then used for the highly dispersed flow regions with small bubble sizes [284–286].

Simulation of the polymer flooding and hydraulic fracturing process

As a further application case for the NCSPH algorithm, the simulation of the so called 'polymer flooding' (also known as 'chemical flooding') process in the context of enhanced oil recovery methods is proposed. During that tertiary phase of the oil production, water is pumped into the geological formation together with a polymeric material, with the aim of pushing the residual crude oil left in the porous rocks ahead and out of the reservoir to the surface. For this purpose, the mobility level of the water-polymer mixture and thus its viscosity has to be adjusted to the higher crude oil viscosity. Since the resulting suspension behaves like a viscoelastic fluid, the corrected NCSPH approach is especially suited for simulating the suspension flow with predominating boundary effects as shown in section 4.2.3. In order to model the displacement of crude oil out of the porous stone, the developed approach may have to be enhanced to include capillary and interface tension forces. Moreover, the approach could be extended to model the hydraulic fracturing process, as the fracture of the elastic, respectively elastic-plastic and brittle rocks by the fluid can be conveniently described within the SPH framework. Possibly needed fracture laws for brittle solids are available in the literature.

Appendices

Appendix A

Grid-based methods

In the following section, the applicability of grid-based methods for the simulation of the morphogenesis of open–porous materials is discussed. The subsequent paragraphs are taken from the science proposal at the Deutsche Forschungsgemeinschaft written by the author for the regarded project [269].

For most engineering applications, the finite element or finite volume methods are widely used. Since the finite element method (FEM) provides greater flexibility to model complex geometries in comparison to the finite volume method (and also as the conventional finite differences method), the following section is narrowed down to finite elements.

Using the FEM, the continuum is divided into a number of discrete elements, which need to be connected by a topological mesh. On that mesh, the finite element interpolation functions are built, ensuring the compatibility of the interpolation. However, in Lagrangian type FEM computations, serious mesh distortion can occur during the simulation of e.g. large material deformations. Thereby, the accuracy of the numerical scheme can be dramatically deteriorated, which can even lead to the termination of the computation. These problems can in parts be alleviated by using adaptive FEM approaches. Nevertheless, additional difficulties are introduced not only through the remeshing procedure, but also through mapping the state variables form the old mesh onto the new one. The latter introduces numerical errors, making frequent remeshing undesirable. A further improvement can be archived by the so called Arbitrary Lagrangian Eulerian (ALE) formulation [15, 16, 287]. By making the mesh independent of the material, the mesh distortion can be minimized. But even with the ALE formulation, very large deformations still lead to numerical errors as it has been reported by Li and coworker. In addition, convective transport leads to spurious oscillations, which need to be stabilized, e.g. by artificial diffusion [36]. From a computational point of view it would be extremely helpful to discretize the continuum only by a number of points or particles, without mesh constraints. According to Li, "... this is the leitmotiv of contemporary particle or meshfree methods" [36].

Besides modeling large deformations, the prediction of the temporal evolution of free surfaces needs to be considered for the simulation of open-porous materials. Various techniques have emerged in grid-based discretization schemes, which can be classified in two approaches, namely interface tracking methods and interface capturing methods. Representative of the former include the front tracking method [288] and the particle tracking scheme Marker-and-Cell [289]. In both approaches, particles without mass are inserted close to the surface and their trajectories are tracked. Additionally a moving mesh can be implemented, which follows the moving surface [290]. The disadvantage of the mentioned techniques lies in the increased computational expense due to the computation of the trajectories of a considerable amount of particles (especially in three dimensions). Furthermore, complex deformation of the surface, like break-up and recombination of interfaces can not be calculated, since the solution is calculated only within the fluid domain [291]. The most frequently used interface capturing methods are the Volume of Fluid (VOF) method [292, 293] as well as the Level Set method [294]. In both methods the surface is extended to a transition zone. In doing so, the interface is smeared, which can lead to less accurate results. For the VOF method a transport equation for the volume fraction has to be solved to track the interface, while in the Level Set Method a distance function is updated periodically. Different from surface tracking techniques, the surface is calculated over the combined multiphase domain. Hence, complex deformations can be solved using very fine grids and small time steps, leading also to higher CPU costs.

As mentioned, the coupled calculation of the two-phase problem, as it is done in the interface capturing methods, enables one to determine also complex surfaces. What seems as an advantage of grid-based methods in conjunction with an appropriate interface capturing scheme, e.g. VOF, turn out to be unnecessary for particle based methods. Using particle methods two approaches are possible. In the first case, both phases are modeled by particles as it is done using the grid-based approaches with interface capturing scheme. In the second approach, the liquid or solid phase is modeled by the particle method. Therewith, the position of the surface is determined by the position of the particles. The gas phase acts as a boundary condition on the free surface. By adequate discretization of the gas phase in the evolving void space, the influence of the gas phase is still included in the calculation. Both approaches are discussed in chapter 3.

Appendix B

Discussion of further particle methods

In the following section, the applicability of further particle methods for the simulation of the morphogenesis of open–porous materials is discussed. The subsequent paragraphs are taken from the science proposal at the Deutsche Forschungsgemeinschaft written by the author for the regarded project [269].

Lattice Boltzmann Method

A originally probabilistic method gaining increasing popularity and capable of solving multiphase flow in complex domains is the so called Lattice Boltzmann Method (LBM). As its successor Lattice Gas Automata, LBM is considered as a mesoscopic particle method [29]. LBM is based on the Boltzmann transport equation, describing the temporal evolution of a density distribution in statistical mechanics with respect to position and momentum [295, 296]. A fluid represented by particle distributions exists on a set of discrete points, which are regularly spaced to form a lattice. At every discrete time step, particles jump, possibly under applied forces, from one lattice node to the next according to their velocity. At the new node, particle collisions occur subsequently, leading to new velocity vectors. Hence, the simulation proceeds in a repetition of propagation and collision steps. In order to mimic macroscopic material properties, the collision rules have to be designed in such a way, that the time-average motion of the particles is consistent with the desired continuum behavior, e.g. the Navier-Stokes equation. Despite of the beauty of deriving the macroscopic behavior from mesoscopic dynamics, the LBM is not used in the present work due to the following reasons: First, the modeling of viscoplastic material behavior is still challenging. According to Ginzburg and Steiner LBM '... needs further investigation for strong viscoplastic regimes where iterative improvement of the effective viscosity approximation may become necessary in collision and/or reconstruction step" [297]. Especially at low Reynolds number flows, the accuracy of LBM and the free surface model is still unsatisfactory for high viscosities [297]. These conditions are of particular interest for the desired applications, the reaction-induced formation of an open-porous material by release of a blowing agent. Regarding that application, the major handicap of LBM lies in the lattice approach. Even if LBM is classified as particle method, it is not a true mesh-free method, since the particle distributions are evaluated on the lattice nodes. Hence, some of the advantages of (Lagrangian) particle methods regarding the application to the applied project mentioned above are lost. E.g. an explicit surface capturing method is needed to model the gas-liquid interactions during the pore forming process. This has been demonstrated by Thuerey, used a VOF-like approach for the description of the liquid-gas interface during the simulation of the foaming of a Newtonian fluid. Therefore, he introduced an additional variable, the volume fraction of the fluid and additional interface lattice cells for dividing fluid and gas lattice cells [33].

Dissipative Particle Dynamics

Another approach for modeling flow of complex fluids on the mesoscopic scale is the so called Dissipative Particle Dynamics (DPD) method, which was initially devised by Hoogerbrugge and Koelman [298]. Similar to Lattice Boltzmann Method, DPD was developed to overcome limitations of the Lattice Gas Automata, in this case lattice artefacts [299]. And like LBM, DPD can be seen as a "coarse graining of molecular dynamics" [21]. A set of point particles, which can be interpreted as a collection of molecules, moves with repulsive conservative, dissipative (to reduce velocity differences between particles) and stochastic forces acting on the other surrounding particles [299]. By adding additional interactions to these particles, non-Newtonian or other complex fluid behavior can be modeled in a simple and versatile way as stated by Espanol. However, several drawbacks are associated with standard DPD. First, it is not possible to relate the continuum parameter viscosity of the fluid with the model parameters used in DPD, despite a correct macroscopic hydrodynamic behavior is resembled [299]. Furthermore, the thermodynamic behavior of the fluid is only governed by the conservative forces and no equation of state can be specified [300]. In addition, the physical length and time scale of the simulation actually simulated are not clear. Based on that, Ellero *et al.* suggested the *thermodynamic consistent* DPD technique, which alleviates some of the mentioned problems [104]. But since thermal fluctuation do not play a role on the physical length scale of the application case at hand, the DPD technique seems not appropriate. Nevertheless, thermodynamic consistent DPD bears striking resemblance with the continuum methods Smoothed Particle Hydrodynamics (SPH) and Moving Particle Semi-Implicit (MPS). For this reason, Espanol *et al.* invented the so called Smoothed Dissipative Particle Dynamics method, which is largely the combination of SPH and DPD [115]. Should it become obvious in future years, that thermal fluctuations play a role in other desired applications, the developed SPH framework can also be extended towards the method developed by Espanol and coworker. Nevertheless, typical applications of DPD are the simulation of multiphase fluid flow through fractures and porous media with complex geometries and wetting behaviors [301], as well as modeling of colloidal and polymeric solutions [21].

Appendix C

SPH discretization from Variational Principles

As stressed in section 2.6 the conservation principle is inherently satisfied if the so called variational principles approach is used to link the divergence operator of the continuity equation with the one of the momentum balance [87]. In the following, the derivation of a variational consistent divergence operator for the stress tensor depending on the chosen representation of the density

$$\frac{D\varrho_i}{Dt} = \varrho_i \sum_j V_j(\boldsymbol{v}_j - \boldsymbol{v}_i)\nabla_j w_{ij}. \qquad (C.1)$$

is sketched. The basic theory and formalism presented in the current paragraph follows the work of Landau and Lifschitz as well as Price and Monaghan [20, 76, 89, 159]. The equations of motion for the particles can be derived from the Hamilton's principle or principle of stationary action. Therewith, the discretization scheme for the momentum equation can be obtained from a Lagrangian L, here for a non–dissipative and compressible fluid, which is simply the difference between the kinetic and internal energy [86, 160, 161]

$$L = \int \varrho \left(\frac{1}{2}\boldsymbol{v}^2 - u(\varrho, s) \right) dV, \qquad (C.2)$$

with u being the specific internal energy, dependent on the fluid density ϱ and specific entropy s. After discretization using the basic SPH kernel interpolation (equation (2.2)), the Lagrangian reads in SPH notation

$$L = \sum_j m_j \left(\frac{1}{2}\boldsymbol{v}_j^2 - u_j(\varrho_j, s_j) \right). \qquad (C.3)$$

The principle of least action, also called Hamilton's principle, requires

$$\delta \int L \, dt = \int \delta L \, dt = \int \left(L(\boldsymbol{x} + \delta\boldsymbol{x}, \boldsymbol{v} + \delta\boldsymbol{v}, t) - L(\boldsymbol{x}, \boldsymbol{v}, t) \right) dt \stackrel{!}{=} 0, \qquad \text{(C.4)}$$

where variations with respect to a small change in particle positions $\delta\boldsymbol{x}_i$ and corresponding variations $\delta\boldsymbol{v}_i$ are considered. The latter are related by

$$\frac{D\delta\boldsymbol{x}_i}{Dt} = \delta\boldsymbol{v}_i \qquad \text{(C.5)}$$

and it is assumed, that the only non–zero variation is $\delta\boldsymbol{x}_i$ [20]. After approximating the Lagrangian to the first order accuracy

$$\int \left(\delta\boldsymbol{x}_i \frac{\partial L}{\partial \boldsymbol{x}_i} + \delta\boldsymbol{v}_i \frac{\partial L}{\partial \boldsymbol{v}_i} \right) dt = 0, \qquad \text{(C.6)}$$

and integration by parts of the second summand, one yields

$$\int_{t_1}^{t_2} \left(\delta\boldsymbol{x}_i \frac{\partial L}{\partial \boldsymbol{x}_i} \right) dt + \left[\delta\boldsymbol{x}_i \frac{\partial L}{\partial \boldsymbol{v}_i} \right]_{t_1}^{t_2} - \int_{t_1}^{t_2} \delta\boldsymbol{x}_i \frac{D}{Dt} \frac{\partial L}{\partial \boldsymbol{v}_i} dt = 0. \qquad \text{(C.7)}$$

By applying the boundary condition, the second term vanishes due to $\delta\boldsymbol{x}(t_1) = \delta\boldsymbol{x}(t_2) \stackrel{!}{=} 0$ and after reorganization, equation (C.4) is transformed to

$$\int \left(\frac{D}{Dt} \frac{\partial L}{\partial \boldsymbol{v}_i} - \frac{\partial L}{\partial \boldsymbol{x}_i} \right) \delta\boldsymbol{x}_i \, dt = 0, \qquad \text{(C.8)}$$

with the term in brackets known as Euler–Lagrange equation [55]. By inserting equation (C.3), one yields

$$\int \left[\frac{D}{Dt}(m_i \boldsymbol{v}_i) + \sum_j m_j \frac{\partial u_j}{\partial \boldsymbol{x}_i} \right] \delta\boldsymbol{x}_i \, dt = 0, \qquad \text{(C.9)}$$

which is further modified to

$$\int \left[\frac{D}{Dt}(m_i \boldsymbol{v}_i) + \sum_j m_j \frac{\partial u_j}{\partial \varrho_j}\bigg|_s \frac{\partial \varrho_j}{\partial \boldsymbol{x}_i} \right] \delta\boldsymbol{x}_i \, dt = 0. \qquad \text{(C.10)}$$

The specific internal energy u can be expressed using the thermodynamic relation

$$du = T \, ds - p \, dv = T \, ds + \frac{p}{\varrho^2} \, d\varrho. \qquad \text{(C.11)}$$

Under the absence of dissipation, respectively a constant specific entropy s, the partial derivative of the internal energy in equation (C.10) can be replaced with

$$\frac{\partial u_i}{\partial \varrho_i}\bigg|_s = \frac{p_i}{\varrho_i^2}. \qquad \text{(C.12)}$$

In the following it is investigated, which type of discretization for the divergence of the stress forces in the momentum equation is variational consistent with the following expression used for discretizing the divergence of the velocity field in the continuity equation. For the discretion of the latter equation (2.39) is used, leading to the expression

$$\frac{D\,\varrho_i}{D\,t} = \varrho_i \sum_j V_j(\boldsymbol{v}_j - \boldsymbol{v}_i)\nabla_j w_{ij}. \qquad (C.13)$$

According to Price and Monaghan, the Lagrangian variation in the density ϱ_i upon virtual displacement results in [76]

$$\delta\varrho_i = \varrho_i \sum_j V_j(\delta\boldsymbol{x}_j - \delta\boldsymbol{x}_i)\nabla_j w_{ij}. \qquad (C.14)$$

Here one has to stress, that virtual displacements are considered as spatial displacements occurring at a fixed time only. Therewith, no dependence on time is considered even for space and time depending functions. Inserting the above equation and equation (C.12) in equation (C.10), one obtains

$$\int \left[\frac{D}{Dt}(m_i\boldsymbol{v}_i) - \sum_j m_j \frac{p_j}{\varrho_j} \sum_k V_k(\delta_{ki} - \delta_{ji})\nabla_k w_{jk} \right] \delta\boldsymbol{x}_i dt = 0. \qquad (C.15)$$

The latter equation is identical with equation (2.75) in section 2.6.

Appendix D

Transformation of coordinates

For the Normalized Corrected SPH (NCSPH) approach, no reconstruction of the truncated interpolation domain is needed as in the conventional SPH approach. In contrast, the respective boundary conditions have to be applied at the boundary particles itself. Therefore, the position and orientation of the interface or boundary needs to be known. E.g. the position of the free surface can be obtained with ease as shown in section 3.2. And the spatial orientation of the latter can be deduced from the unit normal vector to the surface, as shown in section 3.2.2.1. With these informations, the respective boundary conditions can be enforced at the interface. Hence, the stress tensor $\boldsymbol{\sigma}$ is rotated in the coordinate plan of the interface, spanned up by the unit normal vector $\hat{\boldsymbol{n}}$ and the unit tangential vector $\hat{\boldsymbol{t}}$. In two dimensions, the (single) rotation angle α between the two coordinate systems, which is also called Euler angle, can be estimated in various ways [302]. Here α is obtained from the scalar product of the axes of both coordinate systems

$$\alpha = \arccos(\boldsymbol{x} \cdot \hat{\boldsymbol{n}}), \tag{D.1}$$

with $\boldsymbol{x}$ being one unit vector of the reference Cartesian coordinate system or particularly simple by the cross product to give

$$\alpha = \arcsin(-\hat{\boldsymbol{n}}_y), \tag{D.2}$$

respectively

$$\alpha = \arccos(\hat{\boldsymbol{n}}_x). \tag{D.3}$$

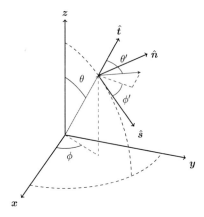

Figure D.1: Coordinate transformation from the Cartesian frame into the normal plane at free surfaces.

One has to stress, that the latter equations are only valid for $\alpha < \pi/2$ and have to be adjusted for larger angles. The rotation matrix is then given as

$$D = \begin{pmatrix} \cos\alpha & -\sin\alpha \\ \sin\alpha & \cos\alpha \end{pmatrix}. \tag{D.4}$$

An arbitrary second order tensor T is then rotated in the normal plane stress tensor $\hat{T}$ as followed

$$\hat{T} = DTD^{-1}. \tag{D.5}$$

The back transformation is simply achieved by

$$T = D^{-1}\hat{T}D. \tag{D.6}$$

In three dimensional calculations, the rotation matrix is more complex and three rotation angles α, β and γ are needed. The rotation angles can be estimated as shown above and the rotation matrix is obtain by multiplication of the three single rotation matrices D^α, D^β and D^γ to give

$$
\begin{aligned}
D &= D^\alpha D^\beta D^\gamma \\
&= \begin{pmatrix} \cos\beta\cos\gamma & \cos\alpha\sin\gamma + \sin\alpha\sin\beta\cos\gamma & \sin\alpha\sin\gamma - \cos\alpha\sin\beta\cos\gamma \\ -\cos\beta\sin\gamma & \cos\alpha\cos\gamma - \sin\alpha\sin\beta\sin\gamma & \sin\alpha\cos\gamma + \cos\alpha\sin\beta\sin\gamma \\ \sin\beta & -\sin\alpha\sin\beta & \cos\alpha\cos\beta \end{pmatrix}
\end{aligned} \tag{D.7}
$$

Appendix E

Corrected second order derivative

In the following paragraph, the derivation of the corrected second order derivative within the Csp method is presented, since second order derivatives are used for the description of dissipative effects in the continuum equations. Starting from equation (2.62), substituting the kernel function $w(\boldsymbol{x} - \boldsymbol{x}', h)$ with the second order derivative $\nabla' \cdot \nabla' w(\boldsymbol{x} - \boldsymbol{x}', h)$ and neglecting higher order terms, one yields the expression for the corrected second order derivatives of particle i as reported by Chen *et al.* to [141]

$$B_{\xi\eta i} f_{\xi i} = \Phi_{\eta i} \tag{E.1}$$

with

$$B_{\xi\eta i} = \zeta \sum_j \frac{m_j}{\rho_j} (x_j^\alpha - x_i^\alpha)(x_j^\beta - x_i^\beta) \frac{\partial w(r_{ij}, h)}{\partial x_j^\gamma \partial x_j^\delta} \tag{E.2}$$

and

$$\Phi_{\eta i} = \sum_j \frac{m_j}{\rho_j} (f_j - f_i) \frac{\partial w(r_{ij}, h)}{\partial x_j^\gamma \partial x_j^\delta} - f_{\alpha i} \sum_j \frac{m_j}{\rho_j} (x_j^\alpha - x_i^\alpha) \frac{\partial w(r_{ij}, h)}{\partial x_j^\gamma \partial x_j^\delta} \tag{E.3}$$

with summation implied over repeated indices. In three dimensions, the rank of the matrix $\boldsymbol{B}$ is six and the corresponding indices are $\xi, \eta = 1, \ldots, 6$ and $\alpha, \beta, \gamma, \delta = 1, \ldots, 3$, as well as $\zeta = 1 - 0.5 \, \delta_{\alpha\beta}$. The index ξ corresponds to $\alpha\beta$ and η corresponds to $\gamma\delta$ as follows: $1 \leftrightarrow 1\,1$, $2 \leftrightarrow 2\,2$, $3 \leftrightarrow 3\,3$, $4 \leftrightarrow 1\,2$, $5 \leftrightarrow 2\,3$, $6 \leftrightarrow 1\,3$. In the second term of the right hand side of vector $\boldsymbol{\Phi}$ (equation (E.3)), the gradient of the field function $\boldsymbol{f}_\alpha$ is included. This feature enables a simple way to directly insert a possible von–Neumann boundary condition. However, in three dimensions, the system of six equations has to be solved for each particle to determine the six independent second order derivatives. This results in increased computational costs, even so the inverse of tensor $\boldsymbol{B}$ with rank six does not have to be computed in each time step, if a symbolic inverse has been determined in advance and the resulting equations are only solved with the actual values in each time step.

The obtained corrected second order derivative is second order complete, therewith approximating constant, linear and parabolic fields exactly [142]. However, since the gradient $\boldsymbol{f}_\alpha$ is used on the right hand side of equation (E.3) and the summation in the second term $\sum_j \frac{m_j}{\rho_j}(\boldsymbol{x}_j^\alpha - \boldsymbol{x}_i^\alpha)w_{\gamma\delta}(r_{ij}, h)$ does not vanish in general, the approximation is only first order complete [142]. The truncation error of the corrected second order derivative, as shown in equation (E.1), is also on the order of h^2 for internal particles and on the order of h for particles on or near boundaries.

The second order derivative scheme which has been evaluated in the present work presented is based on the second derivative of the kernel function, which is in turn corrected using the Cspm approach as shown in section 2.5, respectively equation (E.1). In addition, the accuracy of the conventional Cspm formulation of the second order derivative has been augmented by a renormalization of the kernel function. Due to the renormalization, the derivative is significantly more complex as shown below

$$
\frac{\partial^2 w_{ij}^S}{\partial x_i^\alpha \, \partial x_i^\beta} = \frac{\frac{\partial^2 w_{ij}}{\partial x_i^\alpha \, \partial x_i^\beta}}{\sum_k V_k w_{ik}}
$$
$$
- \frac{\frac{\partial \, w_{ij}}{\partial x_i^\alpha} \sum_k V_k \frac{\partial w_{ik}}{\partial \, x_i^\beta}}{(\sum_k V_k w_{ik})^2}
$$
$$
- \frac{\frac{\partial \, w_{ij}}{\partial x_i^\beta} \sum_k V_k \frac{\partial w_{ik}}{\partial \, x_i^\alpha}}{(\sum_k V_k w_{ik})^2} \qquad \text{(E.4)}
$$
$$
- \frac{w_{ij} \sum_k V_k \frac{\partial^2 w_{ik}}{\partial x_i^\alpha \, \partial x_i^\beta}}{(\sum_k V_k w_{ik})^2}
$$
$$
+ 2\frac{w_{ij} \sum_k V_k \frac{\partial w_{ik}}{\partial \, x_i^\alpha} \sum_k V_k \frac{\partial w_{ik}}{\partial \, x_i^\beta}}{(\sum_k V_k w_{ik})^3}.
$$

With the second order gradient of the renormalized smoothing function, the corrected second order derivative is preserved in the second step using the Cspm approach as shown in equation (E.1). The kernel function of the right hand side of the latter equation can be merged to

$$
\tilde{w}_{ij,\alpha\beta}^S = w_{ij,\alpha\beta}^S - \tilde{w}_{ij,\gamma}^S \sum_k V_k(\boldsymbol{x}_j^\gamma - \boldsymbol{x}_i^\gamma)w_{ij,\alpha\beta}^S, \qquad \text{(E.5)}
$$

with summation implied over repeated indices. It is important to note, that the corrected first order derivative $\tilde{w}_{ij,\gamma}$ shows up in the conditional equation for the corrected second order derivative. The final corrected second order derivative is obtained by solving the linear equation system shown in equation (E.1).

In addition to the presented approach for the discretization of the second order derivative, several other approaches are available as sketched in detail in section 2.1 and 2.2. The choice of the best suited approach is strongly application dependent. If non-homogeneous von Neumann boundary conditions have to be applied at the boundaries, the CSPM based approach presented above is of advantage. The von Neumann condition can be directly inserted in the expression of the second order derivative as in equation (E.5). Using this approach, the simulation of heat transfer at (free) boundaries is simple and straightforward. Therewith, the problem of calculating the surface to volume ratio, e.g. by using the Continuum Surface Force approach, is obsolete [204, 220]. The advantage of the conventional approaches shown in section 2.2 equation (2.55 lies in its simplicity and therefore computational efficiency. However, a larger number of particles is needed within the interaction radius of the kernel function in order to obtain accurate results. Only half of the particle within the interaction domain are needed for the CSPM discretization based on the second derivative of the renormalized kernel function. In doing so, computational advantage of equation (2.55 is at least partly compensated. And by correcting the derivative within the CSPM formalism, the error introduced by a irregular particle distribution is reduced. But the change in the sign of the second order derivative still persists, restricting its application to small cut off radii. As also stated in section 3.2.2, the CSPM second order discretization scheme based on the second order derivative of the kernel function proved to yield less accurate results in comparison to the other investigated approaches used in the verification test case shown in section 4.1.1.

Appendix F

Semi-analytical solution of the evolution of the elliptic fluid drop

In the following, the analytical solution of the pressure evolution inside the fluid drop is sketched for comparison with the SPH solution. The compilation is partly based on the work of Fang *et al.* [228] and a comprehensive derivation of the analytical solution can be found in [67]. The temporal evolution of the semi-minor half-axis a can be analytically obtained as

$$\frac{d\,a}{d\,t} = -aA, \tag{F.1}$$

with A being defined as

$$\frac{d\,A}{d\,t} = \frac{A^2(a^4 - a^2b^2)}{a^4 + a^2b^2}. \tag{F.2}$$

As initial conditions,

$$a(t = 0) = R$$

and

$$A(t = 0) = A_0$$

are used. With known $A(t)$ and $a(t)$, the velocity and pressure fields can be calculated. The expressions for the temporal evolution of the velocity field is obtained as

$$\boldsymbol{v}(\boldsymbol{x}, t) = \begin{pmatrix} -A(t)x \\ +A(t)y \end{pmatrix}$$

and the pressure field is calculated form

$$p^a(\boldsymbol{x}, t) = 0.5\rho \left[\dot{A}(x^2 - y^2) - A^2(x^2 + y^2) - a^2(\dot{A} - A^2) \right]. \tag{F.3}$$

The semi-analytical solution is used to verify the proposed incompressible SPH algorithms in section 4.1.4.

Appendix G

Algorithm for the reconstruction of the pore volume

The scope of the algorithm is to construct a neighbor list for all unoccupied cells which belong to the same pore. With an existing list, the calculation of the pore volume and the respective gas pressure inside the pore can be calculated easily. For this purpose one loop over all unoccupied grid cells is necessary to identify all pores and determine their volumes. The simplified algorithm is sketched below:

```
if i < Number of grid elements then
    if i ! = occupied then
        if i == checked then
            continue
        else
            for j = i + 1 → Number of grid elements do
                if j == checked then
                    continue
                else
                    if (j ! = occupied)&&(j == nearest Neighbor of i) then
                        append i and i's pore neighbors to j's pore neighbor list
                        append j and j's pore neighbors to i's pore neighbor list
                        j ← checked
                        j ← j + 1
                    end if
                end if
            end for
            i ← checked
        end if
    end if
end if
```

G. Algorithm for the reconstruction of the pore volume

Appendix H

Analytical solution of the bubble expansion

In the following, an analytical solution of the expansion of a gaseous bubble in a liquid Newtonian matrix is sketched. The presentation is based on the work of Piesche as well as Bruchon and coworkers [30, 32, 303]. Due to the radial symmetry of the system, the kinematics can be best described in a cylindrical coordinates (r,ϕ). The velocity and pressure field of the liquid matrix depends only on the radial component

$$\boldsymbol{v} = \begin{pmatrix} v_r(r) \\ 0 \end{pmatrix} \tag{H.1}$$

$$p = p(r). \tag{H.2}$$

The incompressibility condition for the fluid phase leads to

$$\nabla \cdot \boldsymbol{v} = \frac{1}{r}\frac{d(rv_r)}{dr} = 0, \quad \text{for } r \geq R \tag{H.3}$$

By separation of the variables with the interface condition $v_r = \frac{dR}{dt}$, the first-order differential equation leads to

$$v_r(r) = \frac{R}{r}\frac{dR}{dt}, \quad \text{for } r \geq R. \tag{H.4}$$

The momentum balance is reduced to a scalar equation due to radial symmetry

$$\frac{d\,\sigma_{rr}}{d\,r} + \frac{1}{r}\left(\sigma_{rr} - \sigma_{\phi\phi}\right) = 0. \tag{H.5}$$

At the gas-liquid interface, the following boundary condition holds:

$$\sigma_{rr}(R) = -p_g \tag{H.6}$$

At the outer boundary of the infinite fluid matrix, the deviatoric stress in radial direction tends towards zero, $\tau_{rr} \to 0$ with $r \to \infty$. This results in

$$\sigma_{rr}(\infty) = -p_{ext}. \tag{H.7}$$

With the definition of the radial velocity $v_r(r)$, as shown in equation (H.4), the deviatoric stress tensor can be expressed as

$$\boldsymbol{\tau} = \eta \frac{\dot{R}R}{r^2} \frac{dR}{dt} \begin{pmatrix} 1 & 0 \\ 0 & -1 \end{pmatrix}. \tag{H.8}$$

Inserting the latter expression in the momentum balance (equation (H.5)) results in

$$\frac{dp}{dr} = 0. \tag{H.9}$$

Therefore, the pressure is constant in the liquid phase

$$p(r) = p_{ext}, \quad \text{for } r > R. \tag{H.10}$$

Based on that, the balance at the interface $r = R$ results in

$$p(R) + \frac{2\eta}{R} \frac{dR}{dt} = -p_g, \tag{H.11}$$

under the assumption of negligible viscosity of the gaseous phase. Hence, the expansion of the gas bubble can be expressed as

$$\frac{1}{R} \frac{dR}{dt} = \frac{p_g - p_{ext}}{2\eta}. \tag{H.12}$$

The expansion velocity is thus governed by the pressure difference. For an ideal gas, the pressure can be formulated as a function of the bubble radius with

$$p_g(t) = \frac{p_g^0 V_g^0}{V_g} = \frac{p_g^0 R^{0^2}}{R^2}. \tag{H.13}$$

Therewith, the temporal evolution of the bubble radius can be simplified to

$$R(t) = R^0 \sqrt{1 + \frac{p^0}{\eta} t}, \tag{H.14}$$

under the assumption of vanishing external pressure $p_{ext} = 0$, which results in an infinite expansion of the gas bubble.

Appendix I

Derivation of balance equations

Thermal energy balance

The following terms have to be considered for the derivation of the thermal energy balance in the Lagrangian form. Starting from an Eulerian perspective

$$\frac{\partial}{\partial t}\left(\varrho(u+\frac{1}{2}v^2)\right) + \frac{\partial}{\partial x^\alpha}\left(\varrho\, v^\alpha\,(u+\frac{1}{2}v^2)\right) =$$
$$= \varrho(\boldsymbol{f}\cdot\boldsymbol{v}) - \frac{\partial(p\,v^\alpha)}{\partial x^\alpha} + \frac{\partial}{\partial x^\beta}(\tau^{\alpha\beta}\,v^\alpha) - \frac{\partial\,q^\alpha}{\partial x^\alpha} + \dot{Q}, \tag{I.1}$$

with the first term on the left hand side representing the energy change within the volume V and the convective flux of energy over the volume surface. On the right hand side, the first term illustrates the change in energy by external volume forces $\boldsymbol{f}$. The second and third term represent the energy change due to pressure forces and deviatoric stress forces (friction), respectively. Last but not least, the energy change due to heat conduction as well as external heat sources is exhibited by the last two terms.

After insertion of the continuity equation (3.3), the latter equation simplifies to

$$\varrho\frac{\partial}{\partial t}(u+\frac{1}{2}v^2) + \varrho\,v^\alpha\frac{\partial}{\partial x^\alpha}(u+\frac{1}{2}v^2) =$$
$$= \varrho(\boldsymbol{f}\cdot\boldsymbol{v}) - \frac{\partial(p\,v^\alpha)}{\partial x^\alpha} + \frac{\partial}{\partial x^\beta}(\tau^{\alpha\beta}\,v^\alpha) - \frac{\partial\,q^\alpha}{\partial x^\alpha} + \dot{Q} \tag{I.2}$$

Multiplying the momentum balance (3.5) with the velocity $\boldsymbol{v}$ and replacing $\boldsymbol{v}d\boldsymbol{v}$ by $d\frac{v^2}{2}$ as well as neglecting any heat sources or sinks, one yields after insertion of the latter

$$\varrho\frac{\partial\,u}{\partial t} + \varrho\,v^\alpha\frac{\partial\,u}{\partial x^\alpha} = -p\frac{\partial v^\alpha}{\partial x^\alpha} + \tau^{\alpha\beta}\frac{\partial v^\alpha}{\partial x^\beta} - \frac{\partial\,q^\alpha}{\partial x^\alpha} + \dot{Q} \tag{I.3}$$

This results in the Lagrangian form

$$\varrho \frac{D\,u}{D\,t} = -p \frac{\partial v^\alpha}{\partial x^\alpha} + \tau^{\alpha\beta} \frac{\partial v^\alpha}{\partial x^\beta} - \frac{\partial\,q^\alpha}{\partial x^\alpha} + \dot{Q} \tag{I.4}$$

After transferring the energy balance in the principle variables temperature T and specific volume v and inserting the component mass balance, the latter equation can be written as

$$\varrho\,c_v \frac{DT}{Dt} + \varrho \left[T \left. \frac{\partial p}{\partial T} \right|_v - p \right] \frac{Dv}{Dt} =$$
$$= -p \frac{\partial v^\alpha}{\partial x^\alpha} - \frac{\partial \dot{q}^\alpha}{\partial x^\alpha} + \tau^{\alpha\beta} \frac{\partial v^\alpha}{\partial x^\beta} - \sum_k MW_k\,u_k \sum_h \nu_{hk}\,r_h. \tag{I.5}$$

The expression $\left. \frac{\partial p}{\partial T} \right|_v$ can be evaluated by

$$\left. \frac{\partial p}{\partial T} \right|_v = -\frac{\alpha}{\beta_T} = -\frac{\left. \frac{\partial V}{\partial T} \right|_p}{\left. \frac{\partial V}{\partial p} \right|_T}. \tag{I.6}$$

For incompressible fluids ($\frac{Dv}{Dt} = 0$) as well as compressible solids, where the thermal expansion ($\alpha = 0$) is neglected, the thermal energy balance is further simplified to

$$\varrho\,c_v \frac{DT}{Dt} = -\frac{\partial \dot{q}^\alpha}{\partial x^\alpha} + \tau^{\alpha\beta} \frac{\partial v^\alpha}{\partial x^\beta} - \sum_k MW_k\,u_k \sum_h \nu_{hk}\,r_h. \tag{I.7}$$

One has to stress, that the isothermal compressibility (β_T) of an elastic solid is the reciprocal of the bulk modulus K [304]. The bulk modulus can be derived from the Poisson ratio ν and Young's modulus E to

$$\frac{1}{\beta_T} = K = \frac{E}{3(1 - 2\nu)}. \tag{I.8}$$

And the remaining $-\varrho p Dv/Dt$ term with

$$\frac{Dv}{Dt} = \frac{1}{\varrho} \frac{\partial v^\alpha}{\partial x^\alpha} \tag{I.9}$$

cancels out with the first term on the right hand side $-p \frac{\partial v^\alpha}{\partial x^\alpha}$ for compressible solids. For an ideal gas phase, the thermal energy balance is formulated as

$$\varrho\,c_v \frac{DT}{Dt} = -p \frac{\partial v^\alpha}{\partial x^\alpha} - \frac{\partial \dot{q}^\alpha}{\partial x^\alpha} + \tau^{\alpha\beta} \frac{\partial v^\alpha}{\partial x^\beta} - \sum_k MW_k\,u_k \sum_h \nu_{hk}\,r_h, \tag{I.10}$$

since the two terms in the square brackets of equation (I.5) cancel each other out.

Component mass balance

Derivation of component mass balance in the Lagrangian form starting form the Eulerian viewpoint

$$\frac{\partial(\varrho\omega_k)}{\partial t} = -\frac{\partial(\varrho\,\omega_k\,\boldsymbol{v}^\alpha)}{\partial\boldsymbol{x}^\alpha} - \frac{\partial j_k^\alpha}{\partial\boldsymbol{x}^\alpha} + MW_k\sum_h \nu_{hk}r_h \tag{I.11}$$

$$\varrho\,\frac{\partial\omega_k}{\partial t} + \omega_k\,\frac{\partial\varrho}{\partial t} = -\boldsymbol{v}^\alpha\varrho\frac{\partial\omega_k}{\partial\boldsymbol{x}^\alpha} - \omega_k\frac{\partial(\varrho\,\boldsymbol{v}^\alpha)}{\partial\boldsymbol{x}^\alpha} - \frac{\partial j_k^\alpha}{\partial\boldsymbol{x}^\alpha} + MW_k\sum_h \nu_{hk}r_h. \tag{I.12}$$

After rearrangement, division with ϱ and introducing the substantial derivative

$$\frac{\partial\omega_k}{\partial t} + \boldsymbol{v}^\alpha\frac{\partial\omega_k}{\partial\boldsymbol{x}^\alpha} = -\frac{\omega_k}{\varrho}\frac{\partial\varrho}{\partial t} - \frac{\omega_k}{\varrho}\frac{\partial(\varrho\,\boldsymbol{v}^\alpha)}{\partial\boldsymbol{x}^\alpha} - \frac{\partial j_k^\alpha}{\partial\boldsymbol{x}^\alpha} + MW_k\sum_h \nu_{hk}r_h \tag{I.13}$$

$$\frac{D\,\omega_k}{D\,t} = -\frac{\omega_k}{\varrho}\frac{\partial\varrho}{\partial t} - \frac{\omega_k}{\varrho}\frac{\partial(\varrho\,\boldsymbol{v}^\alpha)}{\partial\boldsymbol{x}^\alpha} - \frac{\partial j_k^\alpha}{\partial\boldsymbol{x}^\alpha} + MW_k\sum_h \nu_{hk}r_h \tag{I.14}$$

By replacing the partial derivative of the density, one obtains

$$\frac{D\,\omega_k}{D\,t} = \frac{\omega_k}{\varrho}\frac{\partial(\varrho\boldsymbol{v}^\alpha)}{\partial\,\boldsymbol{x}^\alpha} - \frac{\omega_k}{\varrho}\frac{\partial(\varrho\,\boldsymbol{v}^\alpha)}{\partial\boldsymbol{x}^\alpha} - \frac{\partial j_k^\alpha}{\partial\boldsymbol{x}^\alpha} + MW_k\sum_h \nu_{hk}r_h \tag{I.15}$$

$$= -\frac{\partial j_k^\alpha}{\partial\boldsymbol{x}^\alpha} + MW_k\sum_h \nu_{hk}r_h \tag{I.16}$$

References

[1] Ruthven, D. M. *Principles of adsorption and adsorption processes* (John Wiley & Sons, New York) (**1984**).

[2] Keil, F. J. *Modelling of phenomena within catalyst particles.* Chemical Engineering Science, 51:1543–1567 (**1996**).

[3] Strathmann, H. *Trennung von molekularen Mischungen mit Hilfe synthetischer Membranen* (Steinkopf) (**1979**).

[4] Hassan, H. M. and Mumford, C. J. *Mechanisms of drying of skin-forming matrials I., Droplets of materials which gelatinised at high temperature.* Drying technology, 11:1713–1750 (**1993**).

[5] Hassan, H. M. and Mumford, C. J. *Mechanisms of drying of skin-forming materials. II.Droplets of heat sensitive materials.* Drying Technology, 11:1751–1764 (**1993**).

[6] Hassan, H. M. and Mumford, C. J. *Mechanisms of drying of skin-forming materials III. Droplets of natrual products.* Drying Technology, 11:1765–1782 (**1993**).

[7] Krueger, M. *Spruehpolymerisation: Aufbau und Untersuchung von Modellverfahren zur kontinuierlichen Gleichstrom-Spruehpolymerisation.* Ph.D. thesis, Universität Hamburg (**2003**).

[8] Biedasek, S. *Aufbau eines akustischen Levitators zur Durchfuehrung und Online-Verfolgung von Polymerisationen in Einzeltropfen als Modellexperiment für die Spruehpolymerisation.* Ph.D. thesis, University of Hamburg-Harburg (**2009**).

[9] Hammer, J. *Entwicklung zeolithischer Adsorptionsformteile mit thermoplastischen Polymeren als Plastifizier- und Bindemittel.* Ph.D. thesis, Institute of Polymer Technology, University of Stuttgart (**2005**).

[10] Gorbach, A. *Compact Rapid Pressure Swing Adsorption processes - Impact of novel adsorbent monoliths.* Ph.D. thesis, University of Stuttgart (**2005**).

[11] Trefzger, C. *Herstellung zeolithischer Wabenkoerper.* Ph.D. thesis, University of Stuttgart (**2002**).

[12] Fritz, H. G. and Hammer, J. *Aufbereitung zeolitischer Formmassen und ihre Ausformung zu Adsorptionsformteilen.* Chemie Ingenieur Technik, 77:1588–1600 (**2005**).

[13] Fritz, H. G. *personal communication.* 2005.

[14] Hammer, J. *personal communication.* 2005.

[15] Hirt, C. W., Amsden, A. A., and Cook, J. L. *An arbitrary Lagrangian-Eulerian computing for all flow speeds.* Journal of Computational Physics, 14:227–253 (**1974**).

[16] Liu, W. K., Belytschko, T., and Chang, H. *An arbitrary Lagrangian-Eulerian finite element method for path-dependent materials.* Computer Methods in Applied Mechanics and Engineering, 58:227–246 (**1986**).

[17] Belytschko, T., Guo, Y., Liu, W. K., and Xiao, S. P. *A unified stability analysis of meshless particle methods.* International Journal For Numerical Methods In Engineering, 48:1359–1400 (**2000**).

[18] Li, S. and Liu, W. K. *Meshfree Particle Methods* (Springer, Berlin) (**2004**).

[19] Hirt, C. W. and Nichols, B. D. *Volume of Fluid (VOF) Method for the Dynamics of Free Boundaries.* Journal of Computational Physics, 39(1):201–225 (**1981**).

[20] Monaghan, J. J. *Smoothed particle hydrodynamics.* Reports on Progress in Physics, 68(8):1703–1759 (**2005**).

[21] Espanol, P. *Dissipative Particle Dynamics Revisited* (**2002**).

[22] Morris, J. P., Fox, P. J., and Zhu, Y. *Modeling low Reynolds number incompressible flows using SPH*. Journal of Computational Physics, 136(1):214–226 (**1997**).

[23] Takeda, H., Miyama, S. M., and Sekiya, M. *Numerical simulation of viscous flow by Smoothed Particle Hydrodynamic*. Progress of Theoretical Physics, 92:939–960 (**1994**).

[24] Zhang, S., Morita, K., Fukuda, K., and Shirakawa, N. *An improved MPS method for numerical simulations of convective heat transfer problems*. International Journal for Numerical Methods in Fluids, 51(1):31–47 (**2006**).

[25] Premoze, S., Tasdizen, T., Bigler, J., Lefohn, A., and Whitaker, R. T. *Particle-based simulation of fluids*. Computer Graphics Forum, 22(3):401–410 (**2003**).

[26] Swegle, J. W. *Conservation of momentum and tensile instability in particle methods*. Tech. rep., Sandia National Laboratories (**2000**).

[27] Meiburg, E. *Comparision of the Molecular Dynamics method and the Direct Simulation Monte Carlo technique for flows around simple geometries*. Physics of Fluids, 29:3107–3114 (**1986**).

[28] Bird, G. A. *Direct simulation of high-vorticity gas flows*. Physics of Fluids, 30:364–367 (**1987**).

[29] Frisch, U., d'Humieres, D., Hasslacher, B., Lallemand, P., Pomeau, Y., and Rivet, J.-P. *Lattice Gas hydrodynamics in two and three dimensions*. Complex Systems, 1:649–707 (**1987**).

[30] Bruchon, J., Fortin, A., Bousmina, M., and Benmoussa, K. *Direct 2D simulation of small gas bubble clusters: From the expansion step to the equilibrium state*. International Journal for Numerical Methods in Fluids, 54:73–101 (**2007**).

[31] Bikard, J., Bruchon, J., Coupez, T., and Vergnes, B. *Numerical prediction of the foam structure of polymeric materials by direct 3D simulation of their expansion by chemical reaction based on a multidomain method*. Journal of Materials Science, 40(22):5875–5881 (**2005**).

[32] Bruchon, J. and Coupez, T. *A numerical strategy for the direct 3D Simulation of the expansion of bubbles into a molten polymer during a foaming process*. International Journal for Numerical Methods in Fluids, 57:977–1003 (**2008**).

[33] Thuerey, N. *Phyiscally based animation of free surface flows with the Lattice Boltzmann Method*. Ph.D. thesis, University of Erlangen-Nuernberg (**2007**).

[34] Koerner, C., Thies, M., Hofmann, T., Thuerey, N., and Ruede, U. *Lattice Boltzmann Model for free surface flow for modeling foaming*. Journal of Statistical Physics, 121 1-2:179–196 (**2005**).

[35] Shaofan, L. and Liu, W. K. *Meshfree Particle Methods* (Springer Berlin Heidelberg New York) (**2004**).

[36] Li, S. and Liu, W. K. *Meshfree and particle methods and their applications*. Appl. Mech. Rev., 55(1) (**2002**).

[37] Cundall, P. A. and Stack, O. D. L. *A discrete numerical model for granular assemblies*. Geotechnique, 29:47–65 (**1979**).

[38] Grof, Z., Kosek, J., and Marek, M. *Principles of the morphogenesis of polyolefin particles*. Industrial and Engineering Chemistry Research, 44:2389–2404 (**2005**).

[39] Grof, Z., Kosek, J., and Marek, M. *Modelling of the morphogenesis of growing polyolefin particles*. American Institute of Chemical Engineers Journal, 51:2048–2067 (**2005**).

[40] Young, R. J. and Lovell, P. A. *Introduction to polymers* (Chapman & Hall, London, 2nd edition) (**1995**).

[41] Grof, Z. *Meso-scale modelling of the growth and the morphogenesis of polyolefin particles.* Ph.D. thesis, Prague Institute of Chemical Technology, Department of Chemical Engineering (**2004**).

[42] Ledvinkova, B., Keller, F., Kosek, J., and Nieken, U. *Mathematical modeling of the generation of the secondary porous structure in a monolithic adsorbent.* Chemical Engineering Journal, 140:578–585 (**2008**).

[43] Wang, X. L., Zhang, H., Zheng, L. L., and Wei, J. A. *Development of a mesoscopic particle model for synthesis of uranium-ceramic nuclear fuel.* International Journal of Heat and Mass Transfer, 52(21-22):5141–5151 (**2009**).

[44] Liu, G. R. and Liu, M. B. *Smoothed Particle Hydrodynamics: A meshfree particle method* (World Scientific Publishing Company) (**2003**).

[45] Liu, G. R. and Gu, Y. T. *An introduction to meshfree methods and their programming* (Springer Berlin Heidelberg New York) (**2005**).

[46] Hoover, W. G. *Smooth Particle Applied Mechanics: The state of the art* (World Scientific Publishing Company) (**2006**).

[47] Fasshauer, G. E. *Meshfree approximation methods with MATLAB* (World Scientific Publishing Company) (**2007**).

[48] Liu, G. R. *Meshfree Methods: Moving Beyond the Finite Element Method* (CRC Press; 2 edition) (**2009**).

[49] Monaghan, J. J. *Smoothed Particle Hydrodynamics.* Annual Review of Astronomy and Astrophysics, 30:543–574 (**1992**).

[50] Rosswog, S. *Astrophysical smooth particle hydrodynamics.* New Astronomy Reviews, 53(4-6):78–104 (**2009**).

[51] Vignjevic, R. *Review of development of the smooth particle hydrodynamics (SPH) method.* Unpublished Work (**2010**).

[52] Liu, M. B. and Liu, G. R. *Smoothed Particle Hydrodynamics (SPH): an overview and recent developments.* Archives of Computational Methods in Engineering, 17:25–76 (**2010**).

[53] Lucy, L. B. *A numerical approach to the testing of the fission hypothesis.* Astronomical Journal,, 82:1013–1024 (**1977**).

[54] Gingold, R. A. and Monaghan, J. J. *Smoothed particle hydrodynamics - Theory and application to non-spherical stars.* Royal Astronomical Society, Monthly Notices,, 181:375–389 (**1977**).

[55] Gingold, R. A. and Monaghan, J. J. *Kernel estimates as a basis for general particle methods in hydrodynamics.* Journal of Computational Physics, 46(3):429–453 (**1982**).

[56] Monaghan, J. J. *Why particle methods work.* SIAM Journal on Scientific and Statistical Computing, 3:422–433 (**1982**).

[57] Benz, W. *Smoothed Particle Hydrodynamics: The numerical modelling of nonlinear stellar pulsation: Problems and prospects* (Nato ASI Series, Kluwer Academic, Boston, MA) (**1990**).

[58] Hernquist, L. *Some cautionary remarks about smoothed particle hydrodynamics.* Astrophysical Journal, 404:717–722 (**1993**).

[59] Fulk, D. A. *A numerical analysis of smoothed particle hydrodynamics.* Tech. rep., Air Force Institute of Technology (**1994**).

[60] Moris, J. P. *Analysis of Smoothed Particle Hydrodynamics with applications.* Ph.D. thesis, Monash University (**1996**).

[61] Speith, R. *Untersuchung von Smoothed Particle Hydrodynamics anhand astrophysicalischer Beispiele.* Ph.D. thesis, Eberhard-Karls-Universitaet Tuebingen (**1998**).

[62] Laguna, P. *Smoothed particle interpolation.* The Astrophysical Journal, 439:814–821 (**1995**).

[63] Monaghan, J. J. *Particle methods for hydrodynamics.* Computer Physics Reports, 3(2):71–124 (**1985**).

[64] Niedereiter, H. *Quasi-Monte Carlo methods and pseudo-random numbers.* Bulletin of the American Mathematical Society, 84:957–1041 (**1978**).

[65] Wozniakowski, H. *Average case complexity of multivariate integration.* Bulletin of the American Mathematical Society, 24:185–194 (**1991**).

[66] Martin, T. J., Pearce, F. R., and Thomas, P. A. *An owner's guide to Smoothed Particle Hydrodynamics.* arXiv:astro-ph/9310024v1, http://xxx.lanl.gov/abs/astro-ph/9310024v1 (**1993**).

[67] Colagrossi, A. *A meshless Lagrangian method for free surface and interface flows with fragmentation.* Ph.D. thesis, Universita di Roma La Sapienza (**2004**).

[68] Quinlan, N. J., Basa, M., and Lastiwka, M. *Truncation error in mesh-free particle methods.* International Journal for Numerical Methods in Engineering, 66:2064–2085 (**2006**).

[69] Chaniotis, A. K. and Poulikakos, D. *High order interpolation and differentiation using B-splines.* Journal of Computational Physics, 197(1):253–274 (**2004**).

[70] Monaghan, J. J. *Extrapolating B Splines for interpolation.* Journal of Computational Physics, 60:253 (**1985**).

[71] Fulk, D. A. and Quinn, D. W. *An analysis of 1-D smoothed particle hydrodynamics kernels.* Journal of Computational Physics, 126(1):165–180 (**1996**).

[72] Schoenberg, I. J. *Contributions to the problem of approximation of equidistant data by analytic functions. part A. On the problem of smoothing or graduation - a first class of analytic approximation formulae.* Quaterly of Applied Math, IV:45–99 (**1946**).

[73] Monaghan, J. J. and Lattanzio, J. C. *A refined particle method for astrophysical problems.* Astronomy and Astrophysics, 149:135–143 (**1985**).

[74] Morris, J. P. *A study of the stability properties of SPH.* arXiv:astro-ph/9503124v1 (**1995**).

[75] Swegle, J. W., Hicks, D. L., and Attaway, S. W. *Smoothed Particle Hydrodynamics stability analysis.* Journal of Computational Physics, 116(1):123–134 (**1995**).

[76] Price, D. J. *Magnetic fields in Astrophysics.* Ph.D. thesis, University of Cambridge, UK (**2004**).

[77] Liu, M. B., Liu, G. R., and Lam, K. Y. *Constructing smoothing functions in smoothed particle hydrodynamics with applications.* Journal of Computational and Applied Mathematics, 155:263–284 (**2003**).

[78] Gray, J. P., Monaghan, J. J., and Swift, R. P. *SPH elastic dynamics.* Computer Methods in Applied Mechanics and Engineering, 190(49-50):6641–6662 (**2001**).

[79] Graham, D. I. and Hughes, J. P. *Accuracy of SPH viscous flow models.* International Journal for Numerical Methods in Fluids, 56(8):1261–1269 (**2008**).

[80] Belytschko, T., Y. Krongauz, J. D., and Gerlach, C. *On the completeness of Meshfree Particle Methods.* International Journal for Numerical Methods in Engineering, 43:785–819 (**1998**).

[81] Colagrossi, A. and Landrini, M. *Numerical simulation of interfacial flows by smoothed particle hydrodynamics.* Journal of Computational Physics, 191(2):448–475 (**2003**).

[82] Ritchie, B. W. and Thomas, P. A. *Multiphase smoothed-particle hydrodynamics.* Monthly Notices of the Royal Astronomical Society, 323:743–756 (**2001**).

[83] Flebbe, O., Muenzel, S., Herold, H., Riffert, H., and Ruder, H. *Smoothed Partilce*

Hydrodynmics: Physical viscosity and the simulation of accretion disks. The Astrophysical Journal, 431:754–760 (1994).

[84] Hernquist, L. and Katz, N. TREESPH: A unification of SPH with the hierachical tree method. The Astrophysical Journal Supplement Series, 70:419–446 (1989).

[85] Marri, S. and White, S. D. M. Smoothed particle hydrodynamics for galaxy-formation simulations: improved treatments of multiphase gas, of star formation and of supernovae feedback. Monthly Notices of the Royal Astronomical Society, 345:561–574 (2003).

[86] Monaghan, J. J. An introduction to SPH. Computer Physics Communications, 48:89–96 (1988).

[87] Bonet, J. and Lok, T. S. L. Variational and momentum preservation aspects of Smooth Particle Hydrodynamic formulations. Computer Methods in Applied Mechanics and Engineering, 180(1-2):97–115 (1999).

[88] Bonet, J., Kulasegaram, S., Rodriguez-Paz, M. X., and Profit, M. Variational formulation for the smooth particle hydrodynamics (SPH) simulation of fluid and solid problems. Computer Methods in Applied Mechanics and Engineering, 193:1245–1256 (2004).

[89] Price, D. J. and Monaghan, J. J. Smoothed Particle Magnetohydrodynamics - II. Variational principles and varible smoothinglenght terms. Monthly Notices of the Royal Astronomical Society, 248:139–152 (2004).

[90] Chaniotis, A. Remeshed smoothed particle hydrodynamics for the simulation of compressible, viscous, heat conducting, reacting and interfacial flows. Thesis/dissertation, Institute of Technology, ETH Zuerich (2002).

[91] Chaniotis, A., Poulikakos, D., and Koumoutsakos, P. Remeshed Smoothed Particle Hydrodynamics for the simulation of

visous and heat conducting flows. Journal of Computational Physics, 182:67–90 (2002).

[92] Cummins, S. J. and Rudman, M. An SPH projection method. Journal of Computational Physics, 152(2):584–607 (1999).

[93] Watkins, S., Bhattal, A., Francis, N., Turner, J., and Whitworth, A. A new prescription for viscosity in Smoothed Particle Hydrodynamics. Astronomy and Astrophysics Supplement Series, 119:177–187 (1996).

[94] Nugent, S. and Posch, H. A. Liquid drops and surface tension with smoothed particle applied mechanics. Physical Review e, 62(4):4968–4975 (2000).

[95] Jeong, J. H., Jhon, M. S., Halow, J. S., and van Osdol, J. Smoothed particle hydrodynamics: Applications to heat conduction. Computer Physics Communications, 153(1):71–84 (2003).

[96] Meleán, Y., Sigalotti, L. D. G., and Hasmy, A. On the SPH tensile instability in forming viscous liquid drops. Computer Physics Communications, 157:191–200 (2004).

[97] Lopez, H. and Sigalotti, L. D. G. Oscillation of viscous drops with smoothed particle hydrodynamics. Physical Review E, 73(5):051201–1–051201–12 (2006).

[98] Martys, N. S., George, W. L., Chun, B.-W., and Lootens, D. A Smoothed Particle Hydrodynamics-based fluid model with a spatially dependent viscosity: application to flow of a suspension with a non-Newtonian fluid matrix. RHEOLOGICA ACTA, 49(10):1059–1069 (2010).

[99] Jubelgas, M., Springel, V., and Dolag, K. Thermal conduction in cosmological SPH simulations. Monthly Notices of the Royal Astronomical Society, 351:423–435 (2004).

[100] Fathei, R., Fayazbakhsh, M. A., and Manzari, M. T. On discretization of second order derivatives in Smoothed Particle Hydrodynamics. International Journal

for Aerospace and Mechanical Engineering, 3:41–44 (**2009**).

[101] Okuzono, T. *Smoothed-particle method for phase separation in polymer mixtures.* Physical Review E, 56(4):4416–4426 (**1997**).

[102] Okuzono, T. *A numerical study of viscoelastic phase separation in polymer solutions.* Progress in Colloid and Polymer Science, 106:172–174 (**1997**).

[103] Ellero, M., Kroger, M., and Hess, S. *Viscoelastic flows studied by Smoothed Particle dynamics.* Journal of Non-Newtonian Fluid Mechanics, 105(1):35–51 (**2002**).

[104] Ellero, M., Espanol, P., and Flekkoy, E. G. *Thermodynamically consistent fluid particle model for viscoelastic flows.* Physical Review e, 68(4) (**2003**).

[105] Ellero, M. *Smoothed Particle dynamics methods for the simulation of viscoelastic fluids.* Thesis/dissertation, Fakultaet Mathematik und Naturwissenschaften, TU Berlin (**2004**).

[106] Ellero, M. and Tanner, R. I. *SPH simulations of transient viscoelastic flows at low Reynolds number.* Journal of Non-Newtonian Fluid Mechanics, 132(1-3):61–72 (**2005**).

[107] Clavet, S., Beaudoin, P., and Poulin, P. *Particle-based viscoelastic fluid simulation.* In *Symposium on Computer Animation - Eurographics/ACM SIGGRAPH* (**2005**).

[108] Ellero, M., Kroger, M., and Hess, S. *Multiscale modeling of viscoelastic materials containing rigid nonrotating inclusions.* Multiscale Modeling & Simulation, 5(3):759–785 (**2006**).

[109] Fang, J. N., Owens, R. G., Tacher, L., and Parriaux, A. *A numerical study of the SPH method for simulating transient viscoelastic free surface flows.* Journal of Non-Newtonian Fluid Mechanics, 139(1-2):68–84 (**2006**).

[110] Rafiee, A., Manzari, M., and Hosseini, M. *An incompressible SPH method for simulation of unsteady viscoelastic free-surface fows.* International Journal of Non-Linear Mechanics, 42:1210–1223 (**2007**).

[111] Jiang, T., Jie, O. Y., Yang, B. X., and Ren, J. L. *The SPH method for simulating a viscoelastic drop impact and spreading on an inclined plate.* Computational Mechanics, 45(6):573–583 (**2010**).

[112] Vignjevic, R., Campbell, J., and Libersky, L. *A treatment of zero-energy modes in the Smoothed Particle Hydrodynamics method.* Computer Methods in Applied Mechanics and Engineering, 184(1):67–85 (**2000**).

[113] Brookshaw, L. *A method of calculating radiative heat diffusion in particle simulations.* Proceedings of the Astronomical Society of Australia, 6:207–210 (**1985**).

[114] Chikazawa, Y., Koshizuka, S., and Oka, Y. *A particle method for elastic and visco-plastic structures and fluid-structure interactions.* Computational Mechanics, 27(2):97–106 (**2001**).

[115] Espanol, P. and Revenga, M. *Smoothed dissipative particle dynamics.* Physical Review e, 67(2) (**2003**).

[116] Cleary, P. *Modelling confined multi-material heat and mass flows using SPH.* Applied Mathematical Modelling, 22:981–993 (**1998**).

[117] Cleary, P. W. and Monaghan, J. J. *Conduction modelling using Smoothed Particle Hydrodynamics.* Journal of Computational Physics, 148(1):227–264 (**1999**).

[118] Sigalotti, L. D. G., Klapp, J., Sira, E., Meleán, Y., and Hasmy, A. *SPH simulations of time-dependent Poiseuille flow at low Reynolds numbers.* Journal of Computational Physics, 191(2):622–638 (**2003**).

[119] Grenier, N., Antuono, M., Colagrossi, A., Touze, D. L., and Alessandrini, B. *An*

Hamiltonian interface SPH formulation for multi-fluid and free surface flows. Journal of Computational Physics, 228(22):8380–8393 (**2009**).

[120] Fathei, R. and Manzari, M. T. *Error estimation in Smoothed Particle Hydrodynamics and a new scheme for second derivatives.* Computers & Mathematics with Applications, 61:482–498 (**2011**).

[121] Basa, M., Quinlan, N. J., and Lastiwka, M. *Robustness and accuracy of SPH formulations for viscous flow.* International Journal For Numerical Methods In Fluids, 60(10):1127–1148 (**2009**).

[122] Flekkoy, E. G., Wagner, G., and Feder, J. *Hybrid model for combined particle and continuum dynamics.* Europhysics Letters, 52(3):271–276 (**2000**).

[123] Hu, X. Y. and Adams, N. A. *A multiphase SPH method for macroscopic and mesoscopic flows.* Journal of Computational Physics, 213(2):844–861 (**2006**).

[124] Antuono, M., Colagrossi, A., Marrone, S., and Molteni, D. *Free-surface flows solved by means of SPH schemes with numerical diffusive terms.* Computer Physics Communications, 181(3):532–549 (**2010**).

[125] Colagrossi, A., Antuono, M., Souto-Iglesias, A., Touze, D. L., and Izaguirre-Alza, P. *Theoretical analysis of SPH in simulating free-surface viscous flows.* Tech. rep., INSEAN, CESOS, ETSIN, ECN (**2008**).

[126] Schwaiger, H. F. *An implicit corrected SPH formulation for thermal diffusion with linear free surface boundary conditions.* International Journal For Numerical Methods In Engineering, 75:787–800 (**2008**).

[127] Hirsch, C. *Numerical compuation of interal and external flows* (Wiley-Interscience publication) (**1988**).

[128] Lax, P. D. and Richtmyer, R. D. *Survey of the stability of linear finite difference equa-*

tions. Communications on Pure and Applied Mathematics, 9:267–293 (**1956**).

[129] Batra, R. C. and Zhang, G. M. *Analysis of adiabatic shear bands in elasto-thermo-viscoplastic materials by Modified Smoothed Particle Hydrodynamics (MSPH) method.* Journal of Computational Physics, 201(1):172–190 (**2004**).

[130] Belytschko, T., Krongauz, Y., Organ, D., Fleming, M., and Krysl, P. *Meshless methods: an overview and recent developments.* Computer Methods in Applied Mechanics and Engineering, 139:3–47 (**1996**).

[131] Herant, M. *Dirty Tricks for SPH.* In G. Bono and J. C. Miller, eds., *Smoothed Particle Hydrodynamics in Astrophysics,* vol. 65 of *Journal of the Italian Astronomical Society,* 1013–1022 (Societa Astronomica Italiana) (**1994**).

[132] Campbell, P. M. *Some new algorithms for boundary value problems in smoothed particle hydrodynamics.* Tech. rep., DNA Report (**1989**).

[133] DeLeffe, M., LeTouze, D., and Alessandrini, B. *Normal flux method at the boundary for SPH.* Proceedings of the 4th SPERIC workshop, 149 (**2009**).

[134] Marrone, S. *Enhanced SPH modeling of free-surface flows with large deformations.* Ph.D. thesis, University of Rome (**2011**).

[135] Monaghan, J. J. *Simulating free surface flows with SPH.* Journal of Computational Physics, 110:399–406 (**1994**).

[136] Das, A. K. and Das, P. K. *Bubble evolution through submerged orifice using Smoothed Particle Hydrodynamics: Basic formulation and model validation.* Chemical Engineering Science, 64:2281–2290 (**2009**).

[137] Das, A. K. and Das, P. K. *Equilibrium shape and contact angle of sessile drops of different volumes-Computation by SPH and its further improvement by DI.* Chemical Engineering Science, 65(13):4027–4037 (**2010**).

[138] Libersky, L. and Petschek, A. G. *Smoothed Particle Hydrodynamics with Strength of Materials*. Advances in the Free-Lagrange Method Including Contributions on Adaptive Gridding and the Smooth Particle Hydrodynamics Method, Lecture Notes in Physics, 395:248–257 (**1991**).

[139] Yildiz, M., Rook, R. A., and Suleman, A. *SPH with the multiple boundary tangent method*. International Journal for Numerical Methods in Engineering, 77:1416–1438 (**2008**).

[140] Shadloo, M. S., Zainali, A., Sadek, S. H., and Yildiz, M. *Improved Incompressible Smoothed Particle Hydrodynamics method for simulating flow around bluff bodies*. Computer Methods in Applied Mechanics and Engineering, 200:1008–1020 (**2011**).

[141] Chen, J. K., Beraun, J. E., and Carney, T. C. *A corrective smoothed particle method for boundary value problems in heat conduction*. International Journal for Numerical Methods in Engineering, 46(2):231–252 (**1999**).

[142] Chen, J. K., Beraun, J. E., and Jih, C. J. *Completeness of corrective smoothed particle method for linear elastodynamics*. Computational Mechanics, 24(4):273–285 (**1999**).

[143] Chen, J. K. and Beraun, J. E. *A generalized smoothed particle hydrodynamics method for nonlinear dynamic problems*. Computer Methods in Applied Mechanics and Engineering, 190(1-2):225–239 (**2000**).

[144] Chen, J. K., Beraun, J. E., and Jih, C. J. *An improvement for tensile instability in Smoothed Particle Hydrodynamics*. Computational Mechanics, 23:279–287 (**1999**).

[145] Belytschko, T., Lu, Y. Y., and Gu, L. *Element-Free Galerkin methods*. International Journal for Numerical Methods in Engineering, 37:229–254 (**1994**).

[146] Liu, W. L., Jun, S., Li, S., Adee, J., and Belytschko, T. *Reproducing kernel particle methods for structural dynamics*. International Journal for Numerical Methods in Engineering, 38:1655–1679 (**1995**).

[147] Dilts, G. A. *Moving-least-squares-particle hydrodynamics I:. Consistency and stability*. International Journal for Numerical Methods in Engineering, 44:1115Ü–1155 (**1999**).

[148] Dilts, G. A. *Moving least-squares particle hydrodynamics II: conservation and boundaries*. International Journal for Numerical Methods in Engineering, 48:1503Ü–1524 (**2000**).

[149] Atluri, S. *A new Meshless Local Petrov-Galerkin approach in computational mechanics*. Computational Mechanics, 22:117–127 (**2000**).

[150] Liu, G. R. and Gu, Y. T. *A point interpolation method for two-dimensional solids*. International Journal for Numerical Methods in Engineering, 50:937–951 (**2001**).

[151] Johnson, G. R. and Bessel, R. A. *Normalised smoothing functions for SPH impact computations*. Journal for Numerical Method in Engineering, 39:2725–2741 (**1996**).

[152] Zhang, G. M. and Batra, R. C. *Modified Smoothed Particle Hydrodynamics method and its application to transient problems*. Computational Mechanics, 34(2):137–146 (**2004**).

[153] Liu, M. B., Xie, W. P., and Liu, G. R. *Modeling incompressible flows using a finite particle method*. Applied Mathematical Modelling, 29(12):1252–1270 (**2005**).

[154] Shepard, D. *A two-dimensional function for irregularly spaced data*. ACM National Conference (**1968**).

[155] Randles, P. W. and Libersky, L. D. *Smoothed Particle Hydrodynamics: Some recent improvements and applications*. Computer Methods in Applied Mechanics and Engineering, 139:375–408 (**1996**).

[156] Johnson, G. R., Stryk, R. A., and Beissel, S. R. *SPH for high velocity impact computations*. Computers Methods in Applied Mechanics and Engineering, 139:347–373 (**1996**).

[157] Li, S. and Liu, W. K. *Moving Least-square Kernel Galerkin method (II) Fourier Analysis*. Computer Methods in Applied Mechanics and Engineering, 139:159–193 (**1996**).

[158] Liu, W. K., Li, S., and Belytschko, T. *Moving least-square reproducing kernel methods (I) Methodology and convergence*. Computer Methods in Applied Mechanics and Engineering, 143:113–154 (**1997**).

[159] Landau, L. D. and Lifschitz, E. M. *Lehrbuch der theoretischen Physik I - Mechanik* (Harri Deutsch) (**1987**).

[160] Eckart, C. *Variation principles of hydrodynamics*. Physics of Fluids, 3:421–427 (**1960**).

[161] Salmon, R. *Hamiltonian fluid mechanics*. Annual Review of Fluid Mechanics, 20:225–256 (**1988**).

[162] Benz, W. and Asphaug, E. *Impact simulations with fracture .1. method and tests*. Icarus, 107(1):98–116 (**1994**).

[163] Benz, W. and Asphaug, E. *Simulations of brittle solids using Smooth Particle Hydrodynamics*. Computer Physics Communications, 87(1-2):253–265 (**1995**).

[164] Monaghan, J. J. *On the problem of penetration in particle methods*. Journal of Computational Physics, 82(1):1–15 (**1989**).

[165] Koshizuka, S., Tamako, H., and Oka, Y. *A particle method for incompressible viscous flow with fluid fragmentation*. Computational Fluid Dynamics, 4(1):29–46 (**1995**).

[166] Lo, E. Y. M. and Shao, S. D. *Simulation of near-shore solitary wave mechanics by an incompressible SPH method*. Applied Ocean Research, 24(5):275–286 (**2002**).

[167] Gotoh, H. and Sakai, T. *Key issues in the particle method for computation of wave breaking*. Coastal Engineering, 53(2-3):171–179 (**2006**).

[168] Ellero, M., Serrano, M., and Espanol, P. *Incompressible smoothed particle hydrodynamics*. Journal of Computational Physics, 226(2):1731–1752 (**2007**).

[169] Hu, X. Y. and Adams, N. A. *An incompressible multi-phase SPH method*. Journal of Computational Physics, 227(1):264–278 (**2007**).

[170] Hu, X. Y. and Adams, N. A. *A constant-density approach for incompressible multi-phase SPH*. Journal of Computational Physics, 228(6):2082–2091 (**2009**).

[171] Khayyer, A., Gotoh, H., and Shao, S. *Corrected incompressible SPH method for accurate water-surface tracking in breaking waves*. Coastal Engineering, 55:236–250 (**2008**).

[172] Khayyer, A., Gotoh, H., and Shao, S. *Enhanced predictions of wave impact pressure by improved incompressible SPH methods*. Appled Ocean Research, 31(2):111–131 (**2009**).

[173] Issa, R. *Numerical assessment of the Smoothed Particle Hydrodynamics gridless method for incompressible flows and its extension to turbulent flows*. Ph.D. thesis, University of Manchester, Institute of Science and Technology (**2005**).

[174] Lee, E. S. *Truly incompressible approach for computing incompressible flow in SPH and comparisons with the traditional weakly compressible approach*. Ph.D. thesis, University of Manchester (**2007**).

[175] Rafiee, A. and Thiagarajan, K. P. *An SPH projection method for simulating fluid-hypoelastic structure interaction*. Computer Methods in Applied Mechanics and Engineering, 198(33-36):2785–2795 (**2009**).

[176] Noutcheuwa, R. K. and Cwens, R. G. *A new incompressible Smoothed Particle Hydrodynamics-Immersed Boundary Method.* International Journal of Numerical Analysis and Modeling, Series B, 3:126–167 (**2012**).

[177] Ovaysi, S. and Pini, M. *Direct pore-level modeling of incompressible fluid flow in porous media.* Journal of Computational Physics, 229(19):7456–7476 (**2010**).

[178] Lee, E. S., Moulinec, C., Xu, R., Violeau, D., Laurence, D., and Stansby, P. *Comparisons of weakly compressible and truly incompressible algorithms for the SPH mesh free particle method.* Journal of Computational Physics, 227(18):8417–8436 (**2008**).

[179] Batchelor, G. K. *An Introduction to Fluid Dynamics* (Cambridge University Press) (**1967**).

[180] Price, D. J. *Modelling discontinuities and Kelvin-Helmholtz instabilities in SPH.* Journal of Computational Physics, 227(24):10040–10057 (**2008**).

[181] Oechslin, R., Janka, H. T., and Marek, A. *Relativistic neutron star merger simulations with non-zero temperature equations of state, I. Variation of binary parameters and equation of state.* Astronomy and Astrophysics, 467:395–409 (**2007**).

[182] Knapp, C. E. *An Implicit Smoothed Particle Hydrodynamic Code.* Ph.D. thesis, Los Alamos National Laboratory, Universtiy of California (**2000**).

[183] Chorin, A. J. *Numerical solution of Navier-Stokes equations.* Mathematics of Computation, 22(104):745–762 (**1968**).

[184] Temam, R. *Une methode d'approximation des solutions des equations Navier-Stokes.* Bulletin de la Societe Mathematique de France, 98:115–152 (**1968**).

[185] Abdallaha, S. *Numerical solutions for the pressure Poisson equation with Neumann boundary conditions using a non-staggered grid.* Journal of Computational Physics, 70:182–192 (**1987**).

[186] Bell, J., Colella, P., and Glaz, H. *A second order projection method for the incompressible Navier-Stokes equation.* Journal of Computational Physics, 85:257–283 (**1989**).

[187] Johansen, H. and Colella, P. *A Cartesian grid embedded boundary method for Poisson's equation on irregular domains.* Journal of Computational Physics, 147(1):60–85 (**1998**).

[188] Gresho, P. M. and Chan, S. T. *On the theory of semi-implicit projection methods for viscous incompressible flow and its implementation via a finite element method that also introduces a nearly consistent mass matrix.* International Journal for Numerical Methods in Fluids, 11:621–659 (**1990**).

[189] Koshizuka, S., Nobe, A., and Oka, Y. *Numerical analysis of breaking waves using the moving particle semi-implicit method.* International Journal for Numerical Methods in Fluids, 26(7):751–769 (**1998**).

[190] Colin, F., Egli, R., and Lin, F. Y. *Computing a null divergence velocity field using smoothed particle hydrodynamics.* Journal of Computational Physics, 217(2):680–692 (**2006**).

[191] Chorin, A. J. and Marsden, J. E. *A mathematical introduction to fluid mechanics* (Springer Berlin Heidelberg New York) (**2000**).

[192] Merzinger, G. and Wirth, T. *Repetitorium der Hoeheren Mathematik* (Binomi Verlag) (**1999**).

[193] Koshizuka, S. and Oka, Y. *Moving-Particle Semi-Implicit method for fragmentation of incompressible fluid.* Nuclear Science and Engineering, 123:421–434 (**1996**).

[194] Ataie-Ashtiani, B. and Shobeyri, G. *Numerical simulation of landslide impulsive waves*

by incompressible smoothed particle hydrodynamics. International Journal for Numerical
Methods in Fluids, 56(2):209–232 (**2008**).

[195] Ataie-Ashtiani, B., Shobeyri, G., and
Farhadi, L. *Modified incompressible SPH
method for simulating free surface problems.*
Fluid Dynamics Research, 40(9):637–661
(**2008**).

[196] Xu, R., Stansby, P., and Laurence, D. *Accuracy and stability in incompressible SPH
(ISPH) based on the projection method and
a new approach.* Journal of Computational
Physics, 228(18):6703–6725 (**2009**).

[197] Pozorski, J. and Wawrenczuk, A. *SPH
computation of incompressible viscous flows.*
Journal of Theoretical and Applied Mechanics, 40:917 (**2002**).

[198] Schuetz, S. *Modellbildung und Simulation
von Stroemungsvorgaengen.* Tech. rep., Institute of Mechanical Process Engineering
(**2006**).

[199] Bird, R. B., Stewart, W. E., and Lightfoot,
E. N. *Transport Phenomena* (John Wiley
& Sohns, Inc.) (**2002**).

[200] Williams, M. . L., Landel, R. F., and Ferry,
J. D. *The temperature dependence of relaxation mechanisms in amorphous polymers
and other glass-forming liquids.* Journal of
the American Chemical Society, 1955:3701–
3706 (**1955**).

[201] Piesche, M. and Schuetz, S. *Numerische Berechnung mehrphasiger Stroe-
mungen.* Tech. rep., Department of Mechanical Process Engineering, University of
Stuttgart (**2005**).

[202] Kolev, N. I. *Multiphase flow dynamics Vol.
1* (Springer Berlin Heidelberg New York)
(**2005**).

[203] Antoci, C., Gallati, M., and Sibilla, S. *Numerical simulation of fluid-structure interaction by SPH.* Computers & Structures,
85(11-14):879–890 (**2007**).

[204] Brackbill, J. U., Kothe, D. B., and Zemach,
C. *A Continuum method for modeling
surface-tension.* Journal of Computational
Physics, 100(2):335–354 (**1992**).

[205] Morris, J. P. *Simulating surface tension
with Smoothed Particle Hydrodynamics.* International Journal for Numerical Methods
in Fluids, 33(3):333–353 (**2000**).

[206] Cromer, A. *Stable solutions using the Euler approximation.* American Journal of
Physics, 49:455 (**1981**).

[207] Courant, R., Friedrichs, K., and Lewy, H.
*Ueber die partiellen Differenzengleichungen
der mathematischen Physik.* Mathematische
Annalen, 100:32–74 (**1928**).

[208] Xue, S. C., Tanner, R. I., and Phan-Thien,
N. *Numerical modelling of transient vis-
coelastic flows.* Journal of Non-Newtonian
Fluid Mechanics, 123(1):33–58 (**2004**).

[209] Tannehill, J. *Computational fluid mechanics and heat transfer* (Taylor & Francis,
Washington, DC) (**1997**).

[210] Colagrossi, A., Bouscasse, B., Antuono, M.,
and Marrone, S. *Particle packing algorithm
for SPH schemes.* Computer Physics Communications, 183:1642–1653 (**2012**).

[211] Grenier, N., Touzé, D. L., Antonuo, M.,
and Colagrossi, A. *An improved SPH formulation for multi-phase flow simulations.*
In *Proceedings. of 8th International Conference on Hydrodynamics (ICHD, Nantes,
France* (**2008**).

[212] Monaghan, J. J. and Rafiee, A. *A simple SPH Algorithm for multi-fluid flow with
high density ratios.* International Journal
For Numerical Methods In Fluids, 71:573–
561 (**2013**).

[213] Monaghan, J. J. *SPH without a tensile instability.* Journal of Computational Physics,
159(2):290–311 (**2000**).

[214] Russell, P. A. and Abdallah, S. *Dilation-
free solutions for the incompressible flow*

equations on nonstaggered grids. American Institute of Aeronautics and Astronautics Journal, 35:585–586 (**1997**).

[215] Meister, A. *Numerik linearer Gleichungssysteme. Eine Einfuehrung in moderne Verfahren* (Vieweg Friedr. + Sohn Verlag) (**2005**).

[216] Colagrossi, A., Antuono, M., and Touze, D. L. *Theoretical considerations on the free-surface role in the Smoothed Particle Hydrodynamics model.* Physical Review E, 79(5) (**2009**).

[217] Shao, S. D. and Lo, E. Y. M. *Incompressible SPH method for simulating Newtonian and non-Newtonian flows with a free surface.* Advances in Water Resources, 26(7):787–800 (**2003**).

[218] Colagrossi, A., Delorme, L., Cercós-Pita, J., and Souto-Iglesias, A. *Influence of Reynolds number on shallow sloshing flows.* Proceedings of the Third International SPHERIC Workshop, Lausanne, Switzerland. (**2008**).

[219] Ataie-Ashtiani, B. and Farhadi, L. *A stable Moving-Particler Semi-implicit method for free surface flows.* Fluid Dynamics Research, 38(4):241–256 (**2006**).

[220] Saeckel, W., Keller, F., and Nieken, U. *Modeling of spray drying processes using meshfree simulation methods.* Proceedings of the 17th International Drying Symposium, 539–546 (**2010**).

[221] Libersky, L. D., Petschek, A. G., Carney, T. C., Hipp, J. R., and Allahdadi, F. A. *High-strain Lagrangian hydrodynamics - A 3-dimensional SPH code for dynamic material response.* Journal of Computational Physics, 109(1):67–75 (**1993**).

[222] Fathei, R. and Manzari, M. T. *A consistent and fast weakly compressible smoothed particle hydrodynamics with a new wall boundary condition.* International Journal For Numerical Methods In Fluids, 68:905–921 (**2012**).

[223] Vaughan, G. L. *SPH for Fluids.* Tech. rep., Department of Mathematics and Statistics, University of Otago, Dunedin, New Zealand (**2008**).

[224] Khayyer, A. *Improved Particle Methods by Refined Differential Operator Models for Free-Surface Fluid Flows.* Ph.D. thesis, Kyoto University (**2008**).

[225] Lind, S. J., Xu, R., Stansby, P. K., and Rogers, B. D. *Incompressible Smoothed Particle Hydrodynamics for Free-Surface Flows: A Generalised Diffusion-Based Algorithm for Stability and Validations for Impulsive Flows and Propagating Waves.* Journal of Computational Physics, 231:1499–1523 (**2012**).

[226] Oger, G., Doring, M., Alessandrini, B., and Ferrant, P. *An improved SPH method: Towards higher order convergence.* Journal of Computational Physics, 225:1472–1492 (**2007**).

[227] Monaghan, J. J. *SPH and Riemann solvers.* Journal of Computational Physics, 136(2):298–307 (**1997**).

[228] Fang, J., Parriaux, A., Rentschler, M., and Ancey, C. *Improved SPH methods for simulating free surface flows of viscous fluids.* Applied Numerical Mathematics, 59(2):251–271 (**2009**).

[229] Morris, J. P. and Monaghan, J. J. *A switch to reduce SPH viscosity.* Journal of Computational Physics, 136(1):41–50 (**1997**).

[230] Shaw, A. and Roy, D. *Stabilized SPH-based simulations of impact dynamics using acceleration-corrected artificial viscosity.* International Journal of Impact Engineering, 48:98–106 (**2011**).

[231] Martin, J. C. and Moyce, W. J. *An experimental study of the collapse of liquid columns on a rigid horizontal plane.* In *Mathematical and Physical Sciences*, book chapter IV, 312–324 (**1952**).

[232] Touzé, D. L., Colagrossi, A., Colicchio, G., and Greco, M. *A critical investigation of smoothed particle hydrodynamics applied to problems with free-surfaces*. International Journal for Numerical Methods in Fluids, 73:660–691 (**2013**).

[233] Bierbrauer, F., Bollada, P. C., and Phillips, T. N. *A consistent reflected image particle approach to the treatment of boundary conditions in Smoothed Particle Hydrodynamics*. Computer Methods in Applied Mechanics and Engineering, 198(41-44):3400–3410 (**2009**).

[234] Zhu, H. N., Martys, N. S., Ferraris, C., and Kee, D. D. *A numerical study of the flow of Bingham-like fluids in two-dimensional vane and cylinder rheometers using a Smoothed Particle Hydrodynamics (SPH) based method*. Journal of Non-Newtonian Fluid Mechanics, 165(7-8):362–375 (**2010**).

[235] Lehnart, A., Fleissner, F., and Eberhard, P. *An alternative approach to modelling complex Smoothed Particle Hydrodynamics boundaries*. In *Proceedings of the 5th International SPHERIC SPH Workshop, Manchester, Great Britain* (**2010**).

[236] Monaghan, J. J. and Gingold, R. A. *Shock simulation by the particle method SPH*. Journal of Computational Physics, 52:374–389 (**1983**).

[237] Violeau, D. and Issa, R. *Numerical modelling of complex turbulent free-surface flows with the SPH method: an overview*. International Journal for Numerical Methods in Fluids, 53(2):277–304 (**2007**).

[238] Ott, F. *Smoothed Particle Hydrodynamics - Grundlagen und Tests eines speziellen Ansatzes für viskose Wechselwirkungen*. Studien-/Diplomarbeit (**1995**).

[239] Capone, T. *SPH numerical modelling of impulse water waves generated by landslides*. Ph.D. thesis, Sapienza University of Rome (**2009**).

[240] Papanastasiou, T. C. *Flows of materials with yield*. Journal of Rheology, 31:385–404 (**1987**).

[241] Bingham, E. *Fluidity and plasticity* (McGraw-Hill Book Co., New York.) (**1922**).

[242] Piesche, M. *Grundlagen der Stroemungsmechanik*. Tech. rep., Department of Mechanical Process Engineering, University of Stuttgart (**2002**).

[243] Chatzimina, M., Georgiou, G., and Alexandrou, A. *Wall shear rates in circular Couette flow of a Herschel-Bulkley fluid*. Applied Rheology, 129:117–127 (**2005**).

[244] Giesekus, H. *Stressing behavior of simple shear flow as predicted by a new constitutive model for polymer fluids*. Journal of Non-Newtonian Fluid Mechanics, 12:367Ü374 (**1983**).

[245] Burghardt, W. R., Li, J. M., Khomami, B., and B, B. Y. *Uniaxial extensional characterization of a shear thinning fluid using axisymmetric flow birefringence*. Journal of Rheology, 43:147–165 (**1999**).

[246] Crochet, M. J. *Numerical Simulation of Non-Newtonian Flow* (Elsevier Science Publishers, Amsterdam) (**1984**).

[247] Tanner, R. I. *Engineering Rheology* (Clarendon Press, Oxford) (**1988**).

[248] Waters, N. D. and King, M. J. *Unsteady flow of an elastico-viscous liquid* (**1969**).

[249] Yoo, J. Y. and Joseph, D. D. *Hyperbolicity and change of type in the flow of viscoelastic fluids through channels*. Journal of Non-Newtonian Fluid Mechanics, 19:15–41 (**1985**).

[250] Ellero, M., Espanol, P., and Adams, N. A. *Implicit atomistic viscosities in Smoothed Particle Hydrodynamics*. PHYSICAL REVIEW E, 82(4, Part 2):046702–046708 (**2010**).

[251] Hashemi, M. R. and Manzari, M. T. *SPH simulation of interacting solid bodies suspendend in a shear flow of an Oldroyd-B fluid*. Journal of Non-Newtonian Fluid Mechanics, 166:1239–1252 (**2011**).

[252] Palmer, B., Gurumoorthi, V., Tartakovsky, A., and Scheibe, T. *A component-based framework for Smoothed Particle Hydrodynamics simulations of reactive fluid flow in porous media*. International Journal of High Performance Computing Applications, 24(2):228–239 (**2010**).

[253] Ryan, E. M., Tartakovsky, A. M., Recknagle, K. P., Khaleel, M. A., and Amon, C. *Pore-scale modeling of the reactive transport of chromium in the cathode of a solid oxide fuel cell*. JOURNAL OF POWER SOURCES, 196(1, Sp. Iss. SI):287–300 (**2011**).

[254] Tartakovsky, A. M., Meakin, P., Scheibe, T. D., and Wood, B. D. *A smoothed particle hydrodynamics model for reactive transport and mineral precipitation in porous and fractured porous media*. Water Resources Research, 43(5) (**2007**).

[255] Tartakovsky, A. M., Tartakovsky, D. M., Scheibe, T. D., and Meakin, P. *Hybrid simulations of reaction-diffusion systems in porous media*. Siam Journal on Scientific Computing, 30(6):2799–2816 (**2007**).

[256] Tartakovsky, A. M. *Langevin model for reactive transport in porous media*. PHYSICAL REVIEW E, 82(2, Part 2) (**2010**).

[257] Ryan, E. M., Tartakovsky, A. M., and Amon, C. *Pore-scale modeling of competitive adsorption in porous media*. Journal of Contaminant Hydrology, 120-121:56–78 (**2011**).

[258] Neumann, J. V. and Richtmyer, R. D. *A method for the numerical calculation of hydrodynamic shocks*. Journal of Applied Physics, 21:232 (**1950**).

[259] Lattanzio, C., Monaghan, J. J., Pongracic, H., and Schwarz, M. P. *Interstellar cloud collisions*. Monthly Notices of the Royal Astronomical Society, 215:125–147 (**1985**).

[260] Lattanzio, C., Monaghan, J. J., Pongracic, H., and Schwarz, M. P. *Controlling penetration*. SIAM Journal on Scientific and Statistical Computing, 7:591–598 (**1986**).

[261] Guenther, C., Hicks, D. L., and Swegle, J. W. *Conservative smoothing versus artificial viscosity*. Tech. rep., Sandia National Labs., Albuquerque, NM (United States) (**1994**).

[262] Balsara, D. S. *Von Neumann stability analysis of Smoothed Particle Hydrodynamics - suggestions for optimal algorithms*. Journal of Computational Physics, 121:357–372 (**1995**).

[263] Hicks, D. L., Swegle, J. . W., and Attaway, S. W. *Conservative smoothing stabilizes discrete numerical instabilities in SPH material dynamics computation*. Applied Mathematics and Computation, 85:209–226 (**1997**).

[264] Gourma, M. *Towards better understanding of the Smooth Particle Hydrodynamic method*. Ph.D. thesis, Cranfield University (**2003**).

[265] Rabczuk, T., Belytschko, T., and Xiao, S. P. *Stable particle methods based on Lagrangian kernels*. Computer Methods in Applied Mechanics and Engineering, 193(12-14):1035–1063 (**2004**).

[266] Bonet, J. *Recent advances in SPH simulation of fluid and solid dynamics*. In *Meshfree Methods for Partial Differential Equations IV* (**2007**).

[267] Vignjevic, R., Vuyst, T. D., and Campbell, J. C. *A frictionless contact algorithm for meshless methods*. Computer Modeling in Engineering & Sciences, 13(1):35–47 (**2006**).

[268] Vignjevic, R., Vuyst, T. D., and Campbell, J. C. *A frictionless contact algorithm for meshless methods.* ICCES, 3(2):107–112 (**2007**).

[269] Keller, F. and Nieken, U. *Simulation der Morphologieausbildung von offenporigen Materialien.* Tech. rep., DFG-Antrag Ni932/6-1, Institute of Chemical Process Engineering (**2007**).

[270] Monaghan, J. J. and Kocharyan, A. *SPH simulation of multiphase flow.* Computer Physics Communications, 87(1-2):225–235 (**1995**).

[271] Ott, F. and Schnetter, E. *A modified SPH approach for fluids with large density differences.* arXiv:physics/0303112v3 (**2003**).

[272] Liu, M. B., Liu, G. R., and Lam, K. Y. *A one-dimensional meshfree particle formulation for simulating shock waves.* Shock Waves, 13:201–211 (**2003**).

[273] Adami, S., Hu, X. Y., and Adams, N. A. *A conservative SPH method for surfactant dynamics.* Journal of Computational Physics, 229(5):1909–1926 (**2010**).

[274] Reis, T. and Phillips, T. N. *Lattice Boltzmann model for simulating immiscible two-phase flows.* Journal of Physics A: Mathematical and Theoretical, 40:4033–4053 (**2007**).

[275] Sussman, M., Smereka, P., and Osher, S. *A Level-Set approach for computing solutions to the incompressible two-phase flow.* Journal of Computational Physics, 114:146–159 (**1994**).

[276] Das, A. K. and Das, P. K. *Bubble evolution through a submerged orifice using Smoothed Particle Hydrodynamics: Effect of different thermophysical properties.* Industrial and Engineering Chemistry Research, 48 (18):8726–8735 (**2009**).

[277] Das, A. K. *Incorporation of diffuse interface in smoothed particle hydrodynamics: Implementation of the scheme and case studies.*

[278] Feldman, J. and Bonet, J. *Dynamic refinement and boundary contact forces in SPH with applications in fluid flow problems.* International Journal for Numerical Methods in Engineering, 72(3):295–324 (**2007**).

[279] Yoon, H. Y., Koshizuka, S., and Oka, Y. *A particle-gridless hybrid method for incompressible flows.* International Journal for Numerical Methods in Fluids, 30(4):407–424 (**1999**).

[280] Yoon, H. Y., Koshizuka, S., and Oka, Y. *Direct calculation of bubble growth, departure, and rise in nucleate pool boiling.* International Journal of Multiphase Flow, 27(2):277–298 (**2001**).

[281] Woog, T. *Investigation and implementation of free surfaces and surface tension in the model of smoothed particle hydrodynamics.* Master's thesis, University of Stuttgart, Institute of Chemical Process Engineering (**2011**).

[282] Monaghan, J. J. *SPH Compressible Turbulence.* Monthly Notices of the Royal Astronomical Society, 335:843–852 (**2002**).

[283] Springel, V. and Hernquist, L. *Cosmological Smoothed Particle Hydrodynamics simulations: the entropy equation.* Monthly Notices of the Royal Astronomical Society, 333:649–664 (**2002**).

[284] Kuipers, J. A. M. and Swaaij, W. P. M. *Application of Computational Fluid Dynamics to Chemical Reaction Engineering.* Reviews in Chemical Engineering, 13:1–113 (**1997**).

[285] Ishii, M. *Two-Fluid model for two-phase flow.* Multiphase Science and Technology, 5:134–147 (**1987**).

[286] Marschall, H. *Numerical Simulation of Gas-Liquid Reactors with Bubbly Flow using a hybrid Multiphase-CFD approach.* In CFD in Chemical Reaktion Engineering V (**2008**).

[287] Belytschko, T., Liu, W. K., and Moran, B. *Nonlinear Finite Elements for Continua and Structures* (John Wiley and Sons, New York) (**2000**).

[288] Dafermos, C. M. *Polygonal approximations of solutions of the initial value problem for a conservation law.* Journal of Mathematical Analysis and Applications, 38:33–41 (**1972**).

[289] Harlow, F. H. and Welch, J. E. *Numerical Calculation of Time-Dependent Viscous Incompressible Flow of Fluid with Free Surface.* Physics of Fluids, 8(12):2182–& (**1965**).

[290] Hyman, J. M. *Numerical Methods for Tracking Interfaces.* Physica, 12D:396–407 (**1984**).

[291] Greaves, D. *Simulation of interface and free surface flows in a viscous fluid using adapting quadtree grids.* nternational Journal for Numerical Methods in Fluids, 44:1093–1117 (**2004**).

[292] Mashayek, F. and Ashgriz, N. *A hybrid finite-element-volume-of-fluid method for simulating free surface flows and interfaces.* International Journal for Numerical Methods of Fluids, 20:1363Ü1380 (**1995**).

[293] Thompson, E. *Use of the pseudo-concentration to follow creeping viscous flows during transient analysis.* International Journal for Numerical Methods in Fluids, 6:749–761 (**1986**).

[294] Sethian, J. *Level Set Methods. Evolving Interfaces in Geometry, Fluid Mechanics, Computer Vision, and Materials Science* (Cambridge University Press) (**1996**).

[295] McQuarrie, D. A. *Statistical Mechanics* (Harper and Row, N.Y) (**1976**).

[296] He, X. and Luo, L. *Theory of the lattice Boltzmann method: from the Boltzmann equation to the lattice Boltzmann equation.* Physical Review E, 56:6811Ü6817 (**1997**).

[297] Ginzburg, I. and Steiner, K. *A free-surface lattice Boltzmann method for modelling the filling of expanding cavities by Bingham fluids.* Philosophical Transactions of the Royal Society A-Mathematical Physical and Engineering Sciences, 360:453–466 (**2002**).

[298] Hoogerbrugge, P. and Koelman, J. *Simulating Microscopic Hydrodynamics Phenomena with Dissipative Particle Dynamics.* Europhysics Letters, 19:155 (**1992**).

[299] Espanol, P. and Warren, P. *Statistical mechanics of Dissipative Particle Dynamics.* Europhysics Letters, 30 (**1995**).

[300] Groot, R. D. and Warren, P. B. *Dissipative particle dynamics: Bridging the gap between atomistic and mesoscopic simulation.* Journal of Chemical Physics, 107(11):4423–4435 (**1997**).

[301] Liu, M. B., Meakin, P., and Huang, H. *Dissipative particle dynamics with attractive and repulsive particle-particle interactions.* Physics of Fluids, 18(1) (**2006**).

[302] Rose, M. E. *Elementary theory of angular momentum,* (Wiley Subscription Services, Inc., A Wiley Company, New York) (**1957**).

[303] Piesche, M. *Mehrphasenstroemungen.* Tech. rep., Department of Mechanical Process Engineering, University of Stuttgart (**2002**).

[304] Morris, J. W. *Notes on the Thermodynamics of Solids.* Tech. rep., Materials Science and Engineering - University of California, Berkeley (**2007**).